Edible Oils Manual
Second Edition

Editor
Richard A. Della Porta

AOCS
PRESS
Urbana, Illinois

SNACK FOOD
ASSOCIATION
An International Trade Association
Arlington, Virginia

AOCS Press, Urbana, IL 61802

Snack Food Association, Arlington, VA 22209

ISBN 978-1-893997-55-4

Printed in the United States of America.
11 10 09 08 07 5 4 3 2

The paper used in this book is acid-free and falls within the guidelines established to ensure permanence and durability.

Printed with soy-based ink.

Preface

This manual was originally published in 1990 as a layman's guide to understanding the basics of Vegetable Oil Processing, Chemistry, and Applications. I had the pleasure of being a reviewer of the original manuscript, as well as being a user. This updated edition contains the core of the original material — information on this topic has not changed overly much. We have added updated necessary information to reflect current practices and new learnings in the respective areas.

The original contributors did an excellent job and represented their topic areas very well. In this latest update, the contributors provide a fresh look at these same topic areas, making this a working manual for the vegetable oil user of today.

As one of the original reviewers, it has been my privilege to lead the editing effort at the request of the American Oil Chemists' Society in conjunction with the Snack Food Association. I enlisted Robert Reeves of the Institute for Shortening and Edible Oils to review this revised publication as he did for the original. My personal appreciation to Jodey Schonfeld at AOCS for providing the needed impetus to keep me on track with the contributors and for her patience each time the project hit delays. A thank you goes to Gina Clapper at AOCS, for providing me the opportunity to take on this project, and Christopher Melchert at SFA for his role in the coordinated effort between SFA and AOCS to make this project happen.

Together, we have endeavored to produce a manual, which will provide a cornerstone to any program with applications using vegetable oils. I hope the reader finds the effort to be worthwhile.

Richard A. Della Porta

Richard A. Della Porta
Frito-Lay, Inc.

Introduction

Potato, corn, wheat, nuts, and meats are the primary ingredients in salted snacks. These ingredients define many of the products manufactured by members of the Snack Food Association (SFA).

However, another ingredient is just as important in snack manufacturing: edible oils. Edible oils, the cooking medium for the vast majority of snacks, are a key element that plays a major role in determining taste, texture, shelf-life, and influence the nutrient profile of snack foods. The correct purchasing, testing, storage, and processing of these oils are vitally important. The Snack Food Association recognizes the importance of edible oils and the need for a comprehensive reference manual on the proper use and handling of them. We are delighted to provide this important guide to our members.

Special thanks go to the American Oil Chemists' Society, which helped develop this manual and the renowned experts in the edible oil and snack industries who have written and reviewed this manual. The authors and reviewers are listed in the Contents. We believe that using this manual will help our members to ensure the highest possible quality of snack food products.

Sincerely,

James A. McCarthy

Snack Food Association

President and CEO

Contributors

Don Banks is president of Edible Oil Technology, a consulting firm that provides expertise to an international client base for the development, production and use of vegetable oils and derivative products. The firm works with a wide range of clients including seed companies, oil refiners, industrial fryers, snack food companies, bakeries, foodservice organizations, frying equipment manufacturers, and oleochemical companies. Mr. Banks established EOT in 1991 following 28 years of industrial experience.

Thomas G. Crosby has been with Frito-Lay, Inc., as the Advanced Technical Resource in Fats and Oils since 2000. Prior to that time, he was with Bunge for 17 years as a Production/R&D manager. He began his career at Procter & Gamble after receiving his degree in Engineering from Rutgers. His areas of expertise include product formulation, process control, and oil oxidation studies.

Monoj K. Gupta is the president and founder of MG Edible Oil Consulting International, Inc. He founded the company in 1998. He has been in the field of oil technology and food processing for over 39 years, and holds a Master's Degree in Chemical Engineering from the University of Florida, Gainsville, Florida.

Prior to founding MG Edible Oil Consulting International, Inc., Mr. Gupta worked at Frito-Lay, Inc., for 11 years, where he advised the company on oil application for the worldwide operation of PepsiCo.

He is an active member of the American Oil Chemists' Society. He is the past president of the Lipid Oxidation and Quality Division and the South Central Chapter of the Society from 2000–2001. He works closely with the National Sunflower Association; and has served as a consultant for the American Soybean Association in Latin America and the Caribbean.

Richard (Rick) Della Porta is a Principal Scientist, Frito-Lay Research and Development Analytical Laboratories, Plano, Texas, 1984–present. He obtained his B.S. in Environmental Toxicology, from the University of New England, Maine, in 1981, and his M.S. in Chemistry from the University of Texas at Dallas in 1985. Richard has been a member of AOCS since 1991, of AOAC International since 1992, and a manuscript reviewer for *Journal of Agricultural and Food Chemistry* since 2003.

Richard's research interests include: analytical method development in the analysis of lipids and their oxidative reactivity to industrial processes; correlation of oil properties to shelf-life performance and consumer response; application of rapid, non-destructive analysis techniques to provide real-time feedback for process control of continuous flow fryers; and corollary work in lipid soluble component analysis, with emphasis on nutritional applications and implications.

Norman J. Smallwood is the founding and managing partner of The Core Team. Engaging in engineering, production, operations, and executive level management roles with Procter & Gamble, Hunt-Wesson Foods, Lou Ana Foods, A.E. Staley Manufacturing, Zapata Haynie, and The Core Team, he has compiled a record of considerable accomplishment in a broad array of activities.

With the practical application of the behavioral sciences and Jungian psychology, Smallwood is an advocate of developing empowered organizational cultures as a means of achieving business excellence. In this role, he is especially focused on business start-ups, mergers, acquisitions, and renewals.

Contents

Reviewed by Robert Reeves
President, Institute of Shortening
and Edible Oils

Factors Influencing Edible Oil Selection

MONOJ K. GUPTA

MG EDIBLE OIL CONSULTING INTERNATIONAL

INTRODUCTION

In the snack food industry the majority of the oil used is for frying. There are other oil applications in snack foods, such as spraying oil on crackers, spraying oil on fried snack foods for dusting with dry seasoning powder, and making slurry with the oil and dry ingredients for coating fried or baked snack foods. Besides these applications, shortening is used as filler fat for cookies, savory snacks, and shelf-stable baked products.

COMMERCIALLY AVAILABLE OILS

The snack food industry uses commercially available oils sourced throughout the world. There are some exceptions. These are non-commercial oils that are becoming popular for health reasons, but are expensive. These oils are used by the manufacturers to make health claims on specialty products.

Commercial frying oils known around the world are:

1. soybean
2. palm (palmolein—a palm oil fraction)
3. canola (also referred to as edible rapeseed oil)
4. sunflower (includes high and mid-oleic sun flower)
5. corn
6. cottonseed
7. coconut
8. peanut (groundnut)
9. safflower (includes high oleic safflower)
10. sesame seed.

Soybean, palm, canola (rapeseed), and sunflower are the major oils in the world from the standpoint of their volume. Table 1.1 shows the volumes of these oils and the countries that are producing them. Table 1.1 also shows the world production of corn and coconut oil produced. The United States is the primary producer of corn oil. Some of these countries import seeds and crush them; therefore not every country listed in Table 1.1 is growing the specific seeds.

Table 1.2 lists some of the other oilseeds crushed in much lesser volumes around the world. These oils are important for their regional use in frying snack foods.

SELECTION OF OIL FOR FRIED SNACK FOOD

Fats or oils perform multiple functions in the snack food. In fried snack foods, the oil used for frying must have the following attributes:

1. compatible with the base material being fried
2. provide good fried flavor to the product
3. good texture
4. good mouth feel
5. good after taste
6. good storage stability of the product
7. readily available
8. reasonable price
9. good nutritional attributes.

Compatibility between the oil and the base matrix of the fried product is important. The oil should be neutral to the base stock in order for the fried flavor of the product base to come through. This is especially true if the base product is formulated or coated with some formulated coating. On the other hand the oil must be complimentary to the base material and enhance the fried flavor in the product.

COMPATIBILITY

Compatibility between the frying oil and the base material fried is very important. A good example is the use of peanut (groundnut) oil, cottonseed oil, and NuSun oil producing potato chips that have the nutty flavor liked by consumers in the United States. Similarly, corn oil or low linolenic soybean oil produces flavor in tortilla chips that is very desirable to U.S. consumers. Palmolein produces potato chips that do not match the flavor of the cottonseed oil product but has very pleasant flavor and is enjoyed by consumers in many countries. This element of compatibility is influenced by the taste preference

TABLE 1.1 MAJOR WORLD OIL VARIETIES PRODUCED IN CROP YEAR 2004/2005 (IN 1000 METRIC TONS)

Producing Countries	Soybean	Palm	Rapeseed (Canola)	Sunflower	Corn (USA)	Cottonseed
EU-25	2,582	–	5,515	1,737	1,140	–
Canada	299	–	1,256	–	–	–
USA	8,749	–	312	110	–	–
Mexico	680	–	468	–	–	–
Argentina	5,681	–	–	1,525	–	–
Brazil	5,632	–	–	–	–	–
China, PR	5,846	–	4,570	200	–	–
India	922	–	1,846	426	–	–
Japan	625	–	930	–	–	–
South Korea	192	–	–	–	–	–
Taiwan	432	–	–	–	–	–
Other European Countries	–	–	67	756	–	–
Russia	–	–	73	1,779	–	–
Pakistan	–	–	317	–	–	–
Australia	–	–	190	–	–	–
Ukrain	–	–	–	1,179	–	–
Turkey	–	–	–	473	–	–
South Africa	–	–	–	266	–	–
Malaysia	–	15,450	–	–	–	–
Indonesia	–	14,500	–	–	–	–
Niger	–	815	–	–	–	–
Colombia	–	688	–	–	–	–
Thailand	-	720	–	–	–	–
Other Countries	2,401	2,695	319	817	–	–
Total	32,685	33,136	15,863	9,286	1,140	5.03

TABLE 1.2 MINOR WORLD OIL VARIETIES PRODUCED IN CROP YEAR 2004/2005 (IN MILLION METRIC TONS)

Groundnut (Peanut)	Coconut	Sesame
4.49	3.09	0.816

of the consumers at various geographic locations in the world. For example, in Mexico, tortilla chips fried in soybean oil are not as desirable as those fried in palmolein or sunflower oil. Therefore, the compatibility factor must take the regional consumer preference into account.

FRIED FLAVOR

Fried flavor in the snack food is developed by the reaction between the oil, protein, and the carbohydrate (primarily dextrose or other sugars). Perception of the fried flavor can be influenced by the frying oil. This is because the flavor release can vary with different oils for the same snack food. This is also true for the spray oil and the coating oil. There is always some interaction between the oil and the seasoning. Therefore, it is important to know what oil to use in order to achieve the desired fried flavor as well as the flavor from the coating or seasoning dusting on the surface of the product.

TEXTURE

Texture is very characteristic of each individual fried food. Generally, a crisp initial bite is desired in all snack foods. This is achieved through fryer temperature, coating (on coated products), and, in some cases, with the help of higher melting fractions present in the oil.

GOOD MOUTH FEEL AND GOOD AFTERTASTE

Good mouth feel and good aftertaste are contributed by the oil. Mouth feel includes moistness and mouth melt of the product. These attributes are the combined effect of the oil, the base product, and the frying condition.

GOOD STORAGE STABILITY

Good storage stability of the product is vital to the snack food industry. The product in storage can develop oxidized and unpleasant flavor and aftertaste due to the oxidative degradation of the oil. Therefore, oxidative stability of the oil is very critical for its selection. However, one must recognize the fact that the *oxidative stability* of the frying oil and the *flavor stability* of the oil are not synonymous. Oxidative stability of the oil is generally expressed in terms of Active Oxidation Method (AOM) and Oxidative Stability Index (OSI) or Rancimat Method (equivalent to OSI method). However, the flavor stability of the oil depends on the specific oil breakdown compounds formed in the oil during frying as well as the product during storage. Oils with similar AOM values can exhibit very different flavor stability in fried foods.

READY AVAILABILITY AND REASONABLE COST

Ready availability and reasonable cost are important for the selection of the oil as long as the product meets all the necessary standards and meets acceptance of the regional consumers.

GOOD NUTRITIONAL ATTRIBUTES

Finally, good nutritional attributes have become important in the snack food industry. High saturated fat and high *trans* fat are undesirable in all food products including snack foods. Both of these compounds are known to elevate low density lipoprotein (LDL–Bad Cholesterol), depress high density lipoprotein (HDL–Good Cholesterol) and elevate the triglyceride (TG) content in the blood. These factors increase the risk of coronary heart disease.

The U.S. Food and Drug Administration (FDA) has issued a mandate for all food packaged products to indicate the *trans* fat content, as well as other nutritional data. The compliance date was January 1, 2006. This created a great deal of activity in the snack food industry to reduce *trans* fat in all of their products, whether they are fried or baked.

Snack food is eaten for pleasure and indulgence. Therefore, the initial aroma of the product is very important. After the first bite into the food, the consumer looks for the rest of the product characteristics listed previously, as well as texture, mouth feel, and aftertaste.

As the product gets older in storage it begins to lose flavor and texture. Texture loss occurs when the product absorbs moisture through the packaging beyond a certain limit. The flavor of the product deteriorates as the oil in the product gets more and more oxidized while the product in storage. This occurs in

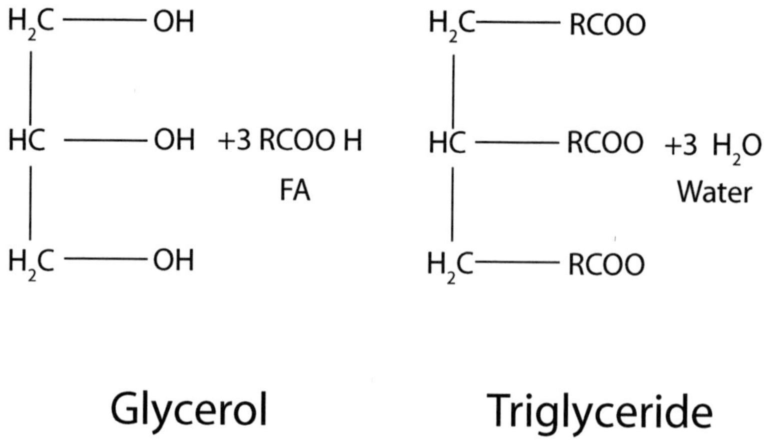

$$\begin{array}{l}
H_2C \text{—} OH \\
| \\
HC \text{—} OH \quad +3\ RCOOH \\
| \qquad\qquad\qquad FA \\
H_2C \text{—} OH
\end{array}
\qquad
\begin{array}{l}
H_2C \text{—} RCOO \\
| \\
HC \text{—} RCOO \quad +3\ H_2O \\
| \qquad\qquad\qquad Water \\
H_2C \text{—} RCOO
\end{array}$$

Glycerol   Triglyceride

FIG. 1.1. STRUCTURE OF TRIGLYCERIDE MOLECULE

addition to the fact that the initial oil quality and the type are very important for the snack food.

Frying oil consists of 96–98% of triglycerides (See Fig. 1.1). The rest consists of some major and minor non-triglyceride components. Some of these components are undesirable for maintaining good stability in the oil. Therefore, these must be reduced to very low levels through the refining process. The other components are beneficial for the oil to maintain good stability. The refining process conditions are maintained to prevent excess damage or loss of these components.

In the frying process, the oil undergoes a series of reactions.

The triglyceride molecules react with water molecules and form free fatty acids (FFA).

The double bonds in the triglycerides react with oxygen forming a series of compounds, some of which impart bad flavor to the product in storage. This reaction process is called autoxidation. These compounds produce poor product flavor during storage.

The heat also produces another group of compounds, called thermal polymers. This happens when the oil is exposed to high heat for a prolonged period without any product going through the fryer. Thermal polymers have a tendency to impart bitter flavor in fried products. This bitterness is different from burnt flavor.

None of the previously mentioned reactions in the oil can be avoided in the frying process. However, oil degradation can be minimized by:

- selecting oil with low level of linolenic acid
- using oils that are naturally more stable in frying application
- taking precautions in the process of frying to minimize the damage to the oil.

COMMERCIALLY AVAILABLE STABLE OILS

There are several commercial oils that are naturally more suitable for the harsh condition of industrial frying and are capable of providing good shelf life to the packaged fried products. The typical examples of these oils are:

- palm oil
- palm olein
- cottonseed
- corn
- peanut (groundnut).

These oils are not from modified composition sources. The natural antioxidants present in these oils make them more suitable for the frying application. Palm oil and palm olein are available in large quantities as commodity oil; and these oils are least expensive in most markets. Cottonseed oil is available at a much lower level worldwide. Corn oil is available especially in the United States and at a volume higher than that of cottonseed oil, but the net volume is much lower than that of palm oil or palm olein. Peanut (groundnut) oil is not available in large quantity and is more expensive than the others.

Discussions on all Commercial Oils

Soybean/Palm/Canola Oil

Two oils are available in large volumes worldwide (see Table 1.1). These are soybean and palm oil. Palm olein is the liquid fraction obtained from palm oil through fractionation process. The third largest oil source is canola (edible rapeseed).

Soybean oil is not suitable for heavy-duty industrial frying because of its high content of linolenic acid. It needs to be hydrogenated to reduce the linolenic acid content to less than 2% in order to use this oil for frying stable packaged snack foods. A similar situation exists with canola oil because of the high linolenic acid content. However, these oils are being used in par-fried frozen products by many U.S. companies. The products are not significantly affected after they are fried and served at the restaurants because the product is stored in the freezer and goes directly to the fryer from the freezer. In addition, the fried product does not develop off flavor because it is served immediately after frying at the restaurant and is consumed within minutes after frying.

Palm oil and palm olein are used in many countries to make stable snack foods. Palm oil may impart some greasy taste or mouth feel in certain products because of its high solids content. Palm olein, on the other hand, can deliver excellent tasting and stable snack foods.

Cottonseed Oil

Cottonseed oil is produced in various countries and consumed locally. The oil contains less than 1% linolenic acid and high level of gamma and delta tocopherols that make the oil very stable for frying. It is also considered to be good for frying because the product exhibits good fried food flavor, which is characterized by some experts and consumers as nutty or buttery in character.

Corn Oil

Corn oil is very good oil for frying shelf-stable snack foods. The oil tends to have slightly lower flavor impact in the freshly fried product. However, the flavor seems to grow with time and eventually has equal or better shelf life than the same product fried in cottonseed oil. Like cottonseed oil, corn oil is also low in linolenic acid. However, the oil is rich in gamma and delta tocopherols.

In addition, the oil also contains derivatives of ferrulic acid, which is a strong antioxidant that provides some extra fry life to it.

Both cottonseed and corn oil contain high levels of linoleic acid, yet they are stable under frying conditions. This can be attributed to the high concentration of the antioxidants as mentioned above.

SUNFLOWER OIL

Classic sunflower oil has essentially the same amount of linoleic acid as cottonseed oil and corn oil. The oil does not produce shelf-stable fried products. This can be attributed to the fact that this oil is rich in alpha tocopherol but has no gamma and delta tocopherols. This makes the oil unsuitable for industrial frying. Mid- and high-oleic sunflower are increasing in availability and have much better frying stability for industrial frying.

PEANUT OIL (GROUNDNUT OIL)

This oil has been popular in many countries for cooking and frying. The oil is high in monounsaturated fatty acid. Besides, the oil is also rich in gamma tocopherol, which makes the oil suitable for industrial frying. Peanut oil and the products fried in the oil have very pleasant flavor. The deterioration of the flavor in the fried food is less observed with this oil unless the oil has been abused in frying. Peanuts have been found to cause allergic reaction in many people. Although fully refined oils are virtually devoid of allergenic protein and thus do not present a risk of allergic reaction, some companies have still refrained from using the oils.

The issue lies more with the cold pressed oil where allergenic protein matter can remain in the oil. Fully processed oil is subjected to very high heat where the pertinacious matter is denatured. Therefore, there is less chance of getting allergic reaction from the oil.

COCONUT OIL

Although this is not a common frying oil, it is used in many countries where coconut is grown in large quantities. The oil imparts some unique nutty flavor character to the fried product. The oil also has a tendency to hydrolyze readily and impart the rancid flavor which is very different from the rancid seed oils.

SAFFLOWER OIL

Safflower oil, like sunflower oil, is not stable under frying condition. However, consumers in Mexico like snack foods fried in safflower as well as in sunflower oil. Safflower oil is produced in extremely limited volume in the world. Varieties of high oleic safflower are now commercially available with good stability for frying applications. Cost is still an issue due to the need for seed stock segregation and refining controls.

SESAME SEED OIL

This is very stable oil for frying because of its unique antioxidants besides the fatty acid composition. The oil produces excellent fried food flavor and provides good storage stability for the fried food. Production of sesame oil is less than 10% of cottonseed annually and mostly located in Asia and the Middle East. Cost per pound is over $0.50, which restricts large scale industrial use.

ANIMAL FATS

It would be appropriate to mention that animal fats were used in the United States for frying and baking until 1987. These fats are still used in some countries for cooking, frying, and baking. Descriptions of the animal fats most commonly used follow.

Beef Tallow

Beef tallow is an excellent frying fat. Straight or partially hydrogenated tallow was used for frying French fries until 1987 in the United States for the unique flavor of the French fries. High saturated fatty acid content of the fat and its impact on CHD (Coronary Heart Disease) resulted in its replacement by hydrogenated soybean, corn, and cottonseed oil in the United States.

Lard

Lard, although fairly high in saturated fatty acid content, is not a stable frying oil due to an insufficient amount of natural antioxidant in it. Lightly hydrogenated, interesterified lard with synthetic antioxidants were used in the United States in the past for frying and baking. It is not used now because of the saturated fatty acid content of the oil.

ADDED ANTIOXIDANTS

Tertiary butyl hydroquinone (TBHQ) at 50–200 ppm concentration is used in frying and baking fats. The net benefit of using this or other synthetic antioxidant in frying as well as some baking fat is questioned by many experts. The author believes that there is some benefit of using synthetic antioxidants in frying or baking fat because they reduce the formation of the free radicals prior to the use of the shortening (fat).

Natural antioxidants such as mixed tocopherols, rosemary extract, etc. are used by some snack food manufacturers. These antioxidants are very costly and one must fully evaluate the cost versus benefit of using them in the snack food. Rosemary can impart its own flavor in the product, thus, limiting its use.

BAKED SNACK FOOD

In baked products, the oil or the shortening impacts several product attributes, such as:

1. flavor
2. texture
3. moistness
4. tenderness (in cakes and pastries)
5. mouth feel (mouth melt)
6. aftertaste
7. storage stability.

Besides the attributes listed, the oil and the shortening must be readily available at reasonable price and have good nutritional traits.

"Flavor," "texture," "moistness," and "mouth feel" of baked products are very different from the fried snacks. Each product has very unique flavor, texture, moistness, and mouth feel. Products such as cookies, crackers, cakes, pastries, cream puffs, etc., are all different from each other. The oil and/or shortening used in these products play a significant role in achieving the desired product attributes. The shortening formula is carefully developed to achieve several attributes, such as:

- batter density
- batter viscosity
- batter stability in the oven
- structural stability of the product as it is taken out of the oven.

"Aftertaste" and "shelf life" of the baked product are greatly affected by the oxidative and flavor stability of the shortening.

"Nutritional" requirements for the baked products have led into a wide range of activity in the area of shortening formulation, oil selection, and the use of fat substitutes. Several modified composition oilseeds have been developed over the past 25 years. Most commonly known oilseeds in this category are:

- high-oleic sunflower
- high-oleic canola
- mid-oleic sunflower (NuSun)
- mid-oleic soybean
- low-linolenic soybean.

The oil produced from these seeds contain low-linolenic (polyunsaturated fatty acid with three double bonds) acid and lower levels of linoleic (polyunsaturated fatty acid with two double bonds) acid and some have lower saturated fatty acid contents. These oils do not require hydrogenation for frying or baking applications. This is what makes these oils very attractive for frying and baking.

TRANS FAT ALTERNATIVES

Natural vegetable oils contain the unsaturated fatty acids in the *cis* form. *Trans* fatty acids are formed during hydrogenation of the vegetable oil. *Trans* fatty acids are unsaturated fatty acids except that they have somewhat different chemical structure (see Fig. 1.2). Diets high in these fatty acids have been found to: (1) increase the bad cholesterol, Low Density Lipoprotein (LDL), (2) lower the level of good cholesterol, High Density Lipoprotein (HDL), and (3) increase the level of triglyceride in human blood serum. All of these factors increase the risk for coronary heart disease (CHD). Modified composition oils mentioned above are being used by several food manufacturers to replace partially hydrogenated oils. The other alternatives that have been found effective to reduce or eliminate *trans* fat are:

1. use of special blends of emulsifiers in certain baked products.
2. use of interesterified shortening where fully saturated oil and liquid

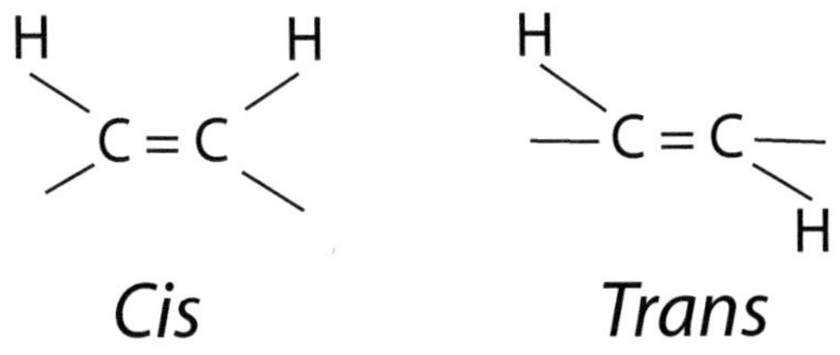

FIG. 1.2 *CIS* VERSUS *TRANS* DOUBLE BOND

  oil are allowed to react together to exchange the fatty acids between the two. The resultant shortening functions very similar to the hydrogenated shortening in baking and some frying applications.

 3. use of palm oil, palmolein and palm stearine along with other stable seed oils to make shortenings for baking.

Unfortunately, the efforts made in steps 1 and 2 have produced shortenings that demonstrate somewhat lower flavor (storage) stability compared to the hydrogenated counterparts. This adds additional challenges to the food product manufacturers. In some cases, the reduced stability in the interesterified products arises from the use of liquid soybean oil, which is not a very stable oil because of its high linolenic acid content. Some oil manufacturers have incorporated antioxidants but this has not produced the same degree of stability as the hydrogenated shortening. This would be alleviated if some of the modified composition oils were used instead of liquid soybean oil. However, the supply of these modified composition oils is very limited. In addition, these oils are very costly and they are not available in large volumes at the moment. The annual need for frying oil in the United States is over 3 billion pounds, whereas the total production of the modified composition oil in the country is barely 1 billion pounds. On the other hand, nearly 12 billion pounds of baking and margarine fats are used in the United States every year. At present, this demand cannot be fully met without the incorporation of some amount of palm oil in the shortening formula or the use of less stable seed oils and sacrificing some shelf life of the finished product.

2 Edible Oil Purchasing

Monoj K. Gupta

MG Edible Oil Consulting International

Purchasing edible oil for food manufacturing is a multifaceted task. It deals with the purchase of oil with good performance at the least delivered cost to the manufacturing plants. There are two kinds of operation where oil is to be purchased for business. They are:

- restaurants
- food processors manufacturing shelf-stable products.

In a restaurant, purchasing oil is less complicated. It is dictated by supply and cost as long as the quality of the finished product is satisfactory. This is because:

1. The fried product does not need to meet a specific shelf life because it is sold immediately or shortly after its preparation.
2. There is less competition compared to the food manufacturers (however, the food preparation in some restaurants might be preferred over the others by the clients).
3. Restaurants operate under very tight net profit margin. Therefore, like all other ingredients, the oil must be purchased at lowest possible cost.

Despite the cost constraints the restaurants must follow certain guidelines regarding oil purchasing as outlined below:

1. The food prepared at the restaurant must have no objectionable flavor or taste.
2. The oil must have long fry-life.
3. The oil must be obtained from a reputable broker.
4. The oil received must be relatively fresh.
5. The broker must have good facility for oil storage and must practice good stock rotation.

Food manufacturer oil procurement is different from the purchase for restaurants and also is somewhat more complicated.

In a corporation, where purchasing is a separate group or department, the latter has to satisfy the management from:

- the product development department
- the manufacturing operation
- their own department
- the corporate profit center.

This makes oil procurement challenging. This challenge may be unique to certain food sectors. For example, a large bakery will have very different needs as compared to a large fried snack food manufacturer. However, some common threads exist between the various businesses which are listed below:

1. understanding company's needs
2. peaks and valleys of the demand for oil required by the company
3. familiarity with the oil supply sources
4. establishing contact with the suppliers
5. keep abreast of the domestic as well as the world supply of oils
6. meeting the company's budget for oil.

Understanding Company's Needs

Have a clear definition of the company's needs in terms of the (1) type of oil, and (2) the volume of oil needed by the company. Large snack food manufacturers need different types of oil because of the unique nature of the product. This adds complexity to the oil purchasers' shopping cart.

Peaks and Valleys of Demand

This is dictated by the sales promotion and seasonal use of the product. For example, during the Memorial Day weekend, Independence Day, start of the football season, etc., the demand of salty snack products goes up. Similarly, the demand for baked products goes up during the Christmas and the Easter holidays. It requires a certain number of weeks to fill up the product supply pipeline to meet the demand. In order to make the product, the company must have a secured oil supply (along with the other ingredients) that must start flowing into the manufacturing plants just in time for the production. Besides these peak demand periods, there may be company promotions on "price-off" or "two for one" sale, etc. This requires certain amount of coordination among the sales, marketing, manufacturing, and oil procurement.

It is difficult to suggest the volume of oil that a company should purchase or contract in advance. In general, any large oil user needs to write a contract with the suppliers that would include the oil price, delivery time, and location.

The buyer may run the risk of running short on oil or may face a large surplus if the company's plans change. However, the change in a company's marketing plan does not occur suddenly. Generally, there is ample time for the oil purchaser to take the required action in such cases. The oil purchaser can sell the oil in the open market when possible. This is done by many corporations and is found profitable to the company in an inflationary market. Normally the oil supplier does not like to buy back the contract unless the supplier finds it to be profitable for their own business.

FAMILIARITY WITH THE SUPPLY SOURCE

There are two supply sources of oils for business. They are:

1. brokerage houses
2. oil manufacturers.

Restaurants and food services typically use the brokerage houses because of the nature of their business. The restaurants or food services need to buy oils or shortening in packages that are suitable for this type of operation. Most oil manufacturers sell their packaged products through the brokers.

The large oil users generally deal directly with the oil manufacturers. However, there are some exceptions. Some oil manufacturers deal only through the brokers. In that event even the large oil users go through the broker designated by the oil manufacturers.

The oil procurement person has to know the supply source, name of the contact person, location of the supplier, etc. There should be a good business relationship between the oil procurement person of the company and the sales and technical support personnel at the supplier. The following guidelines are found to be helpful for oil procurement personnel:

- Know the business volume of the supplier. Establish the production capability of the supplier. Obtain the business pattern and practices of the supplier through professional services (e.g. Dunn & Bradstreet or other services).

- Multiple production plants are essential in order for them to deliver oil at the purchaser's production plants at least cost (at various parts of the country). This also provides the protection against supply interruptions from one of the plants due to any reason.

- Know the technical capability of the supplier, such as their research and development (R&D) and technical services.

- Using their own company's experts, establish the supplier's operating practices regarding quality assurance

- The supplier must have the capability to produce more than only one type of oil. This helps oil procurement to have one-stop shopping when the price is agreeable.

- Oil procurement must also seek different oil suppliers in order to broaden the supply base. This can be used as leverage for obtaining the best price from them.

- With the help of the sales and marketing departments, learn about the seasonal demand because of the promotions at different times during the year. This pattern could be established from the experience over a few years of operation.

- Be familiar with the world oil production and supply. The price of oilseeds and crude oil are determined by the worldwide supply and demand. This kind of information is available through the publications of the United States Department of Agriculture (USDA), Chicago Board of Trading (CBOT) and other private organizations.

Until recently, soybean oil was the number one oil in the world in terms of production. Argentina, Brazil, and the United States are the largest soybean producers. However, palm oil production has exceeded that of soybean oil. The major players in palm oil production are Malaysia and Indonesia. The oil producing countries and the volume of oil produced in the 2004–2005 crop year are shown in Table 1.1 in Chapter 1 "Factors Influencing Edible Oil Selection."

ESTABLISHING CONTRACT ON OIL PRICING

This is an area that requires a considerable amount of knowledge about the supply of oil and the potential demand from the market The oil purchasing personnel use several terms that might be foreign to the rest of the company, e.g.:

- the CBOT or the Chicago Board of Trade
- the Basis points
- cash price — this is quite straightforward.

CBOT

Chicago Board of Trade deals in commodity pricing and futures. This is used by many corporate oil purchasers. They write contracts with oil processors specifying the quantity of oil they want to purchase, the margin over the CBOT price they want to pay, and over what time period they would receive the oil. This works well for large oil users.

THE BASIS POINT

This method uses the difference between the cash and the future prices as quoted by the CBOT. The price may also include the cost of refining and even delivery cost if it is written in the contract.

This can be somewhat risky because the future price may go up and then the company will be required to pay higher price for the oil.

CASH PRICE

This is the actual price paid for the oil at the time of purchase. This is how most small users purchase oil.

There is another method that is used by some buyers who believe that either crude oil price (CBOT) and the basis point will drop on the day of purchase. They write a contract which is known as "Prices Date of Shipment" (PDS). This can work well when the oil price is on the declining slope.

The purchase price is negotiated between the buyer and the seller. Based on the volume of business the seller may agree to sell the refined oil in the United States at an agreed margin over the Chicago Board of Trade (CBOT) crude price at the time of negotiation.

Usually, the broker accepts the contracts from oil buyers over the telephone and without a signature on the dotted line as long the broker is assured that the order is firm.

DOMESTIC AND WORLD SUPPLY OF OILS

This information is for the oil procurement personnel that deals with large volumes of oil every year. The person should be able to foresee the oil supply from the world oil production and supply reports and write the contracts with oil suppliers accordingly. This requires some experience in the vegetable oil commodity market. One needs to stay in tune with the weather conditions in the oil-producing countries, the oil supply, demand, and any potential crisis that might affect the oil supply due to the climatic, economic, or political situation.

MEETING COMPANY'S OIL PROCUREMENT BUDGET

Every company should establish a budget figure for oil for every year. This is done by combining the historical oil usage, the projected increase in oil usage, pricing increase, delivery cost to the plant, etc.

It is essential for oil procurement personnel to make sure that the budget cost for the oil is not exceeded because this would have a severe impact on product pricing and the corporate profitability.

OVERALL COMMENTS ON OIL PURCHASING STRATEGY

Purchasing oil by the large user requires a complete knowledge of the market. Astute oil buyers even have been known to make money for the company from time to time by using the right purchasing strategy.

3 Chemical Composition of Fats and Oils

DON BANKS

EDIBLE OIL TECHNOLOGIES

It is important for snack manufacturers to gain a basic knowledge of the chemical composition of fats and oils in order to understand oil quality and oil breakdown during storage, handling, and usage.

Edible oils are classified as triglycerides. A triglyceride consists of three molecules of fatty acid bonded to a molecule of glycerol as shown in Fig. 3.1. Glycerol is commonly referred to as the "backbone" of a triglyceride. Triglycerides are also called triacylglycerols, which is a more accurate technical term and "acyl" describes the structure of the chemical bonding between fatty acids and glycerol.

FIGURE 3.1. MIXED TRIGLYCERIDE

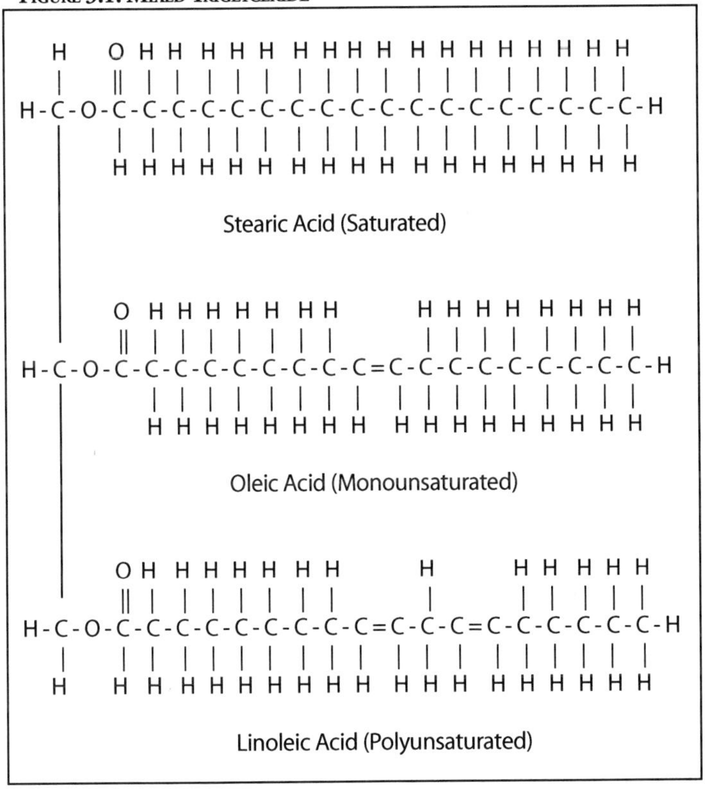

Stearic Acid (Saturated)

Oleic Acid (Monounsaturated)

Linoleic Acid (Polyunsaturated)

Most edible oils contain the same types of fatty acids but in varying percentages, which accounts for differences in both chemical and physical properties. Table 3.1 shows the distribution of fatty acids of a number of common oils. The numbers following the names serve as a shorthand designation indicating the number of carbons and the number of double bonds in the fatty acid. For example, Linoleic, $C_{18:2}$ signifies that this fatty acid contains eighteen carbon atoms and two carbon-to-carbon double bonds. Position of the double bonds and orientation are also indicated by the preceding letters and numbers; $c,t,9,12$-18:2 refers to a *cis* bond at the 9 carbon and a *trans* bond at the 12 carbon of the linoleic acid.

The physical and chemical characteristics of an oil, including melting point and oxidative stability, directly reflect its fatty acid composition. The arrangement of fatty acids on the triglyceride backbone is also a factor but will not be included in this discussion.

TABLE 3.1. % FATTY ACID COMPOSITION OF SELECTED FATS AND OILS

Fatty Acid	Soy	Cotton	Corn	Peanut	Canola	Sun	Coconut	Palm	Palmolein	High Oleic Sunflower	Lard
Capric, $C_{8:0}$							8				
Capric, $C_{10:0}$							6				
Lauric, $C_{12:0}$							48				0.3
Myristic, $C_{14:0}$	0.1	1	0.1	0.1	0.1	0.1	18	1	1	4	1.7
Palmitic, $C_{16:0}$	11	23	11	11.5	4.3	7	9	46	40	4	26.2
Stearic, $C_{18:0}$	3.6	3	2	3.1	1.7	4.2	2	4	3	80	13.5
Oleic, $C_{18:1}$	24.7	18	27	48	59.1	19.5	7	37	42	10	45.2
Linoleic, $C_{18:2}$	53.5	54	58.5	31	22.6	68	2	10	12	0.1	9.5
Linolenic, $C_{18:3}$	6.4	0.4	0.5	1	8.2	0.9		1	1	0.5	0.4
Arachidic, $C_{20:0}$	0.3	0.2	0.3	1.3	0.5	0.3		1	1	1	0.2
Gadoleic, $C_{20:1}$					2						0.7
Behenic, $C_{22:0}$				3							
Erucic, $C_{22:1}$					0.9						

The most common fatty acids contained in edible oils are palmitic ($C_{16:0}$), stearic ($C_{18:0}$), oleic ($C_{18:1}$), and linoleic ($C_{18:2}$). Palmitic and stearic are saturated fatty acids, oleic is monounsaturated, and linoleic is polyunsaturated. Linolenic ($C_{18:3}$) is an "essential" (see Table 3.1) polyunsaturated fatty acid that is present at ≤1% or less in most oils but it is a significant component of both soybean and canola oil.

The term "saturated" is used to describe a fatty acid that contains the maximum number of hydrogen atoms chemically possible. Removing one hydrogen atom from each of two adjacent carbons allows for a carbon-to-carbon double bond to be formed, and the result is a monounsaturated fatty acid. Fatty acids that contain two or more carbon-carbon double bonds are classified as "polyunsaturated."

Saturated fatty acids are very resistant to oxidation and they have high melting points. For example stearic acid has a melting point of 157°F. By removing two hydrogen atoms from stearic acid to form oleic acid, a monounsaturated fatty acid, the melting point is reduced to 61°F. Incorporating a second double bond to form linoleic, the melting point is reduced to 20°F, and by including a third double bond to form linolenic, the melting is reduced point to 9°F.

The reason that saturated fatty acids have high melting points is that they have a relative linear structure that allows for a close or tight association between molecules in the crystalline form. Incorporating a carbon-to-carbon double bond creates an angle or a bend in the fatty acid carbon chain that limits association in the crystalline form and reduces the melting point.

Looking at the fatty acid composition of oils as shown in Table 3.1 and considering the melting point of individual fatty acids, it is not difficult to understand that palm oil melts in the range of 105–108°F, while corn, soy, and sunflower melt at temperatures below 32°F.

In addition to lowering the melting point, a double bond causes a fatty acid to be susceptible to oxidation. Oxidation does not occur at the double bond directly, but at the adjacent carbons in the fatty acid chain. Oleic acid has two susceptible sites, one on each side of the double bond, but with just one double bond, the susceptibility to oxidation is relatively modest. Incorporating a second double bond to form linolenic acid provides three sites susceptible to oxidation, but as can be seen in Fig. 3.1, one of the sites is between two double bonds, which significantly increases susceptibility to oxidation. Incorporating a third double bond to form linolenic acid introduces a second site between double bonds, further increasing susceptibility to oxidation.

The relative rates of autoxidation for oleic, linoleic, and linolenic are on the order of 1:12:25. Analyzing oil and determining the relative percentages of each fatty acid provides good insight to its oxidative stability; however, it should be noted that oxidation of edible oils is a complex process, involving a number of factors that go beyond the current discussion.

When oils are subjected to oxidation, peroxides (i.e. hydroperoxides) are formed that then undergo further reaction to form aldehydes, ketones, short-chain fatty acids, and other secondary oxidation products (i.e. degradation products). Peroxides per se do not have objectionable aromas, rather it is the accumulation of the secondary oxidation products that leads to objectionable aromas and eventually rancidity.

Again looking at fatty acid composition data in Table 3.1 it is readily apparent that palm oil with over 60% saturates and high oleic sunflower oil with about 80% oleic acid are both quite resistant to oxidation. It is also apparent that canola and soybean oils have limited oxidative stability, due in large part to their relatively high concentrations of linolenic acid.

Partial hydrogenation has regularly been used to reduce polyunsaturates and increase oxidative stability and/or incorporate structural stability in the manufacture of frying oil, shortenings, and margarine. To improve the oxidative stability of soybean and canola oils for frying, hydrogenation conditions are controlled to minimize linolenic, reduce linoleic, and increase oleic with minimal increase in stearic. However, during hydrogenation, some *trans* fatty isomers (*trans* fats) are formed, which has raised nutritional issues that will be addressed later in this chapter.

Hydrogenation of edible oils is the major source of *trans* fat is the U.S. diet, but it is not the only source. Ruminate animals naturally hydrogenate oils and produce *trans* fats as they digest their food. As a result, beef, dairy, and lamb products account for some 20% of *trans* fat in the U.S. diet.

The issue of *trans* fat is being addressed at the refinery level by developing methods to reduce or prevent the formation of *trans* isomers. The methods include blending, interesterification and low-*trans* hydrogenation. The issue is also being addressed by seed breeders who have developed new varieties of soybeans, canola, sunflower, and other oilseed crops to yield oils that do not require hydrogenation. Some of the new oils are listed in Table 3.2. Some relatively stable oils currently in the marketplace (e.g. cottonseed, corn, palm) may be used in certain products without additional processing.

From a nutritional standpoint, two polyunsaturated fatty acids—linoleic and linolenic—are especially important. Each of these fatty acids serves as a precursor for the production of specific prostaglandin hormones in the human body. They are classified as "essential" fatty acids because the human body cannot synthesize them, so, like vitamins, they have to be obtained from the food we eat.

There are also nutritional considerations concerning the fatty acid composition of oils and their effect on blood cholesterol. For example, among the saturated fatty acids; lauric, myristic, and palmitic acids are generally considered to raise both LDL (bad cholesterol) and total serum cholesterol. Stearic and oleic acids are generally considered to have a neutral effect and polyunsaturated fatty acids (e.g. linoleic, linolenic) have long been recognized for lowering blood cholesterol.

Data from studies conducted on *trans* fats (i.e. elaidic, $C_{18:1}$) have been inconsistent, showing the same, greater or less impact on cholesterol as compared to saturated fats. Additionally, some studies have shown that adding significant amounts of *trans* fat to a base diet can slightly decrease HDL (good cholesterol), while other studies have not shown any change in HDL. Further testing is needed to resolve the issue; however, the FDA requires labeling of *trans* fat as of January 1, 2006.

In considering options to reduce or eliminate *trans* fat from products, it is noteworthy that *trans* fat, on average, accounts for 2–4% of calories in the U.S. diet as compared to 14–15% for saturated fats; and health organizations have advised against increasing saturated fat to avoid *trans*.

TABLE 3.2. %FATTY ACID COMPOSITION: NEW MODIFIED OILS

Fatty Acid	Low-Lin Soybean	Mid-Oleic Soybean	Low-Lin Canola	Mid-Oleic Canola	Mid-Oleic Sunflower
Myristic, C14:0	0.1	0.1	0.1	0.1	0.1
Palmitic, C:16:0	11	11	4	4	4
Stearic, C18:0	4	4	2	2	2.7
Oleic, C18:1	25	50 – 70	22	65 – 75	50 – 65
Linoleic, C:18:2	57	12 – 27	64	12 – 22	27 – 42
Linolenic, C:18:3	< 3	< 3	4	3	0.4
Arachidic, C20:0	0.3	0.3	0.5	0.5	0.3
Gadoleic, C:20:1			2	2	
Behenic, C22:0					
Erucic, C22:1			0.9	0.9	

OTHER COMPONENTS OF EDIBLE OILS

In addition to triglycerides, most edible oils contain small amounts of phosphatides, sterols, fatty alcohols, waxes, pigments, tocopherols, monoglycerides, and other materials. During oil refining some of these materials, such as phosphatides are removed to the extent possible and others, such as tocopherols, are only partially removed to retain desirable traits.

The most common pigments found in frying oils are chlorophylls and carotenoids. Chlorophylls give oil a greenish tinge but the color can be masked when darker pigments are present. Carotenoids impart a reddish-orange to dark brown color, depending on concentration. Cottonseed oil contains a pigment called gossypol that darkens when heated. Most of these pigments are significantly reduced during processing to produce RBD (refined, bleached, and deodorized) oil.

Tocopherols are naturally occurring vitamin E-active compounds that have some antioxidant properties at low concentrations but can become prooxidants at higher concentrations. When either using added tocopherols or controlling the natural content during processing, an optimum level should be established to insure beneficial usage.

Free fatty acid (FFA) is commonly monitored during snack production as a general measure of oil quality. In a continuous fryer, the FFA should increase gradually and then reach an equilibrium value and remain relatively constant throughout production. However, if problems develop the FFA can rapidly increase and reach unacceptable levels.

Reducing the FFA of used oil, other than by refinery processing, is challenging. Methods for treating high FFA oil are available and most can readily improve oil appearance, but restoring quality is more difficult. When assessing methods for treating high FFA oil, testing should be conducted to verify that oil quality can be recovered and that the benefits extend to the quality and shelf life of the finished products.

Oil Processing for the Production of Snack Foods

Norman J. Smallwood

The Core Team (Partner)

The characteristics of crude edible oils are highly variable with respect to quality and functionality. Several processing operations are required to produce finished oil that meets the demands for snack food quality, taste, and shelf life.

Oil processing includes the removal of many undesirable (nontriglyceride) components and, in some cases, modification of the triglyceride molecules. Each processing step is designed to achieve a specific level of oil quality and/or functionality.

If under-processed oil is passed on to the next step in the processing sequence, the finished product will not meet final specifications for snack food application. The consequences of over-processing oil are to increase the production cost and reduce the product quality and tocopherol content due to the additional heat stress.

Edible Oil Processing

Specific Objectives

Using the applicable steps shown in Fig. 4.1, crude oil is processed to remove undesirable components, enhance quality, and achieve the expected functionality for snack food production. The triglyceride decomposition products of hydrolysis and oxidation must be essentially removed. Phosphatides must be reduced to a trace amount (<2 ppm Phosphorus [P]). Metals, herbicides, and pesticides must also be removed. With some exceptions, reduction of the color level is an aesthetic consideration. Acceptable oil color levels differ by snack food application and are established accordingly.

Because of the health and antioxidant benefits, maximum tocopherol retention is desired. For the typical processing regimen, about 33% of the tocopherol content is removed.

FIGURE 4.1. EDIBLE OIL PROCESSING FLOW DIAGRAM

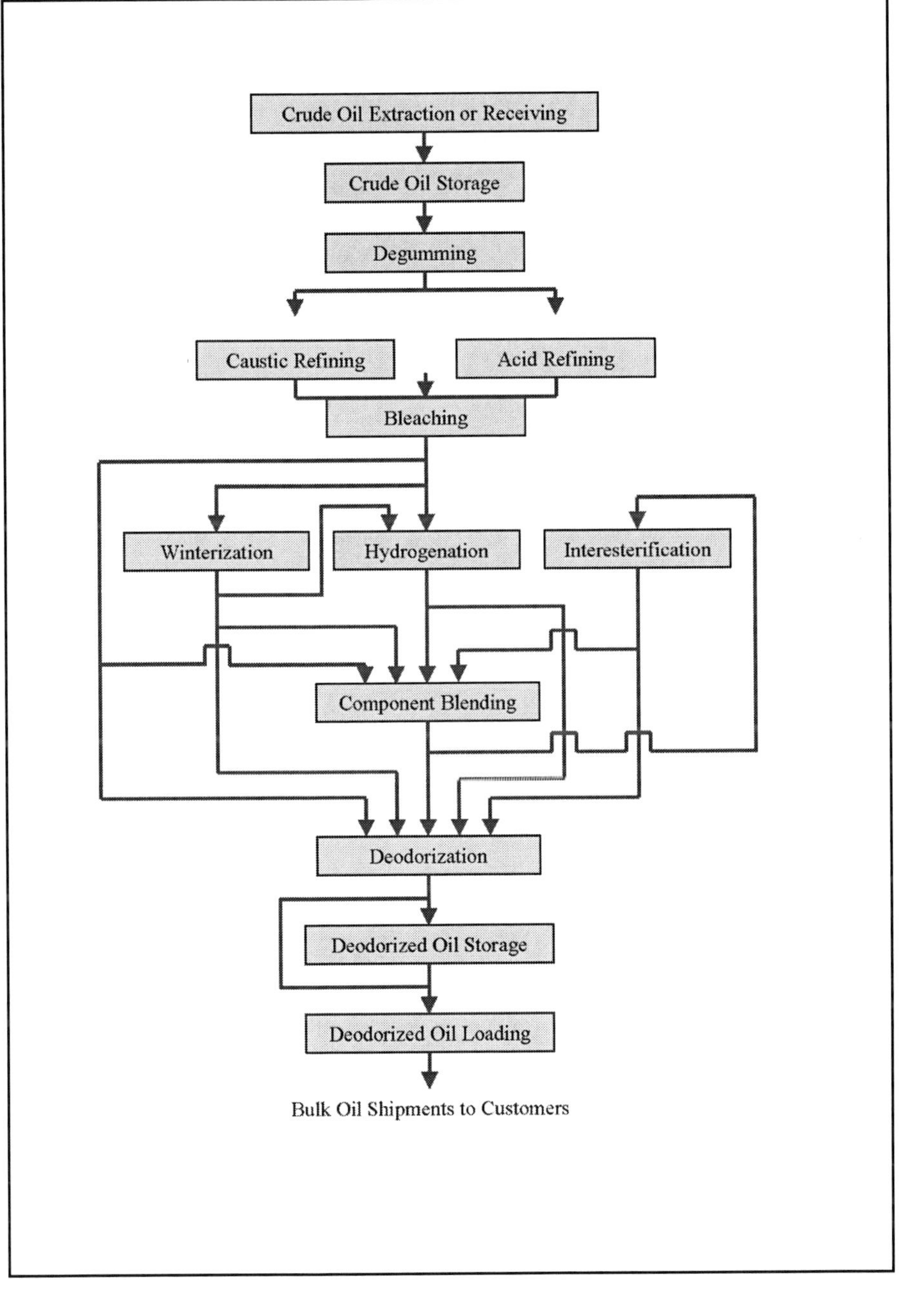

To minimize triglyceride degradation from hydrolysis and oxidation and maximize tocopherol retention, heat stress must be minimized in the processing sequence. Heat stress is a time-temperature consideration.

CRUDE OIL RECEIVING AND UNLOADING

The first potential risk to crude oil quality lies in transportation to the processing facility. The tank car, truck, barge, or other vessel used for shipping must be free of internal contaminants and should protect the crude oil from external contamination. Upon receipt at the processing plant, the oil is sampled and analyzed before unloading. Unloading equipment (fittings, hoses, pumps, and pipelines) must be protected to prevent contamination. If heating is required to melt the oil for unloading, it should be done slowly and evenly to avoid triglyceride degradation.

CRUDE OIL STORAGE

Storage of crude oil prior to processing is required to build sufficient inventory for efficient operation and to maximize yields. Storage tanks must be constructed to provide proper isolation and protection of the oil. Each storage tank should be equipped with effective agitation to provide homogeneous oil feed to processing. A representative sample of the crude oil should be analyzed before the start of each processing run to establish the basis for optimum processing conditions.

Storage time should be as short as possible. The maximum allowable storage time is a function of the crude oil stability and the ambient (storage) temperature.

DEGUMMING

Degumming is frequently the first processing step for crude oils such as soybean and canola that contain a relatively high level of water-hydratable phosphatides. Water is injected into the crude oil feed stream in proportion to the quantity of phosphatides. After sufficient hold time in a surge tank, the process stream is fed to centrifuges for separation of the hydrated phosphatides (gums) from the oil stream.

CAUSTIC REFINING

Removal of free fatty acids, phosphatides, protein meal, glycerol, carbohydrates, resins, and metals is achieved by caustic refining. Caustic soda (sodium hydroxide) is fed in the proper strength and quantity to react with the free fatty acids and phosphatides to form soapstock and hydrated gums. Soapstock and the other impurities are separated from the oil by centrifuges.

ACID REFINING

An alternative to caustic refining is the process of adding a weak acid to the degummed oil for hydration and subsequent removal of the remaining phosphatides to a level of about 5 ppm P. With this process, there is no removal of free fatty acids. The free fatty acid removal is accomplished later in the deodorization process. Compared to caustic refining, the merits of this process include improving yield performance and eliminating the soapstock disposal problems.

Developed in the 1950s, the A.E. Staley Manufacturing Co. used acid refining (acetic anhydride) for over 30 years. In recent years, AGP has developed, used, and promoted acid refining (citric acid). Several edible oil refiners throughout the world have used phosphoric acid in acid refining.

BLEACHING

A common misconception about bleaching is that the primary purpose is to reduce the oil color. In addition to color reduction, it is important for removing oxidation products, residual soap from caustic refining, and metals. Also, the phosphatide level is further reduced. The process involves adding natural or activated clay to the oil stream and providing several minutes of contact between the clay and oil at the appropriate temperature and pressure. The oil-soluble nontriglyceride components are adsorbed on the surface of the clay particles. The clay is then removed along with the adsorbed components from the oil stream by filtration. Adsorption is an equilibrium phenomenon. Thus, complete removal by adsorption is not possible.

HYDROGENATION

Hydrogenation is the process of linking hydrogen to double-bonded carbon atoms in unsaturated fatty acids. By saturating (eliminating) a certain amount of these double bonds, hydrogenation increases the stability of the oil for frying and raises the melting point. For example, if a processor needs to adjust an oil to achieve better oxidative stability for frying or wants to establish a precise melting point for an oil to meet application functional requirements, hydrogenation is often the answer. In the preparation of oils for frying snack foods, hydrogenation is usually applied to reduce the polyunsaturation of soybean and canola oils.

Both soybean and canola oils contain relatively high levels of fatty acid chains containing three double bonds (linolenic fatty acid chains), about 7 and 10%, respectively. To achieve acceptable product fry and shelf life, the linolenic fatty acids must be reduced to less than 3.5% by selective hydrogenation.

The hydrogenation process is usually done in a batch-type reaction vessel equipped with an agitator and a cooling coil. Bleached oil, heated to the proper reaction temperature, is charged to the reaction vessel with a specified type and amount of catalyst. Hydrogen gas is introduced to the vessel through a sparge ring located near the bottom. The oil hydrogenation continues until the desired endpoint is achieved. Because hydrogenation reaction produces heat, cooling water is circulated through the reactor-vessel cooling coil to maintain the specified temperature. After the reaction is complete, the oil-catalyst mixture is pumped from the reactor to a filter for catalyst removal.

Formation of triglyceride fatty acid *trans* isomers is one of the consequences of light to moderate levels of oil hydrogenation. This geometric isomer (change in the fatty acid chain geometry at a double bond position) results in an increase in the melting point for the triglyceride molecule. For fully hydrogenated (completely saturated) fats, the carbon double bonds in the fatty acid chains are essentially eliminated and are practically free of *trans* isomers. While conditions can be established to minimize *trans* isomer formation, complete elimination is not a practical capability for lightly to moderately hydrogenated oils.

With the concern that *trans* isomers pose a health risk, alternatives to hydrogenation are now getting significant attention in the edible oil processing industry. Interesterification is gaining prominence as a substitute for hydrogenation in some product applications.

INTERESTERIFICATION

Edible oil interesterification involves the rearrangement of fatty acid chains among the triglyceride molecules. To accomplish interesterification, a catalyst is required along with the appropriate physical conditions (temperature and pressure). With sodium methylate as the most often used catalyst, chemical interesterification has had some use in edible fats and oils processing for many years. For example, lard has been interesterified to improve functionality in baking applications for many years. In addition, interesterification to produce margarine oils has had some application in Europe for at least 40 years.

By interesterifying a hard oil (highly saturated fat with negligible *trans* isomers) and a *trans-* isomer-free soft oil in the proper proportions, the desired functionality can be achieved for some applications. While not providing the same range of product functionality as hydrogenation, interesterification can be an alternative in some cases.

The advantage of chemical interesterification is that all three fatty acid chains are involved in the random rearrangement. This provides a wider range of functionality possibilities. This advantage is significantly offset by the need to wash and bleach chemically interesterified oils to remove the reaction residues (primarily soap) and the dark brown color.

Enzymatic interesterification has recently become a viable alternative to the chemical process. The enzyme is attached to silica particles positioned as a bed in a column for continuous interesterification. With the enzymatic process, no discoloration of the oil occurs and there are no residues left to remove. However, only the one- and three-position fatty acid chains are randomly interchanged in the enzymatic process. The two-position (middle) chain is not interchanged. Consequently, the range of end-product functionality for enzymatic interesterification is not as robust as compared to the chemical process.

WINTERIZATION

Winterization is a process to separate the saturated fraction (solid fat) from the unsaturated fraction (liquid oil) in an oil. Winterization is commonly used to fractionate palm and cottonseed oils for the end products of olein (the liquid phase) and stearin (the solid phase). The olein fraction is commonly used for salad or frying oil. The stearin fraction is typically used in formulating margarine and baking shortening. The process involves cooling oil at a controlled rate to a temperature low enough to solidify (crystallize) the saturated triglyceride fraction. After the crystal formation is complete, the gently agitated oil is transferred through a filter to separate the liquid and solid portions. While winterization is a practical and often used fractionation technique, it is not highly efficient and is limited in triglyceride separation selectivity.

OIL BLENDING

Because different edible oils possess a wide variety of characteristics, snack food manufacturers frequently specify oil blends for their needs. Edible oil processors produce an array of oils that may include refined and bleached, hydrogenated, winterized, and interesterified. To produce a specific blend requested by a snack food manufacturer, the designated stocks are mixed in precise proportions to meet the finished product specifications.

DEODORIZATION

Deodorization, the final step of the oil processing sequence, is used to remove volatile components such as free fatty acids, glycerol, oxidation products, low molecular weight sterols, herbicides, and pesticides. The oil is heated to a temperature in the range of 450–485°F at about three millimeters mercury (3 mm Hg) pressure. Under these conditions, the nontriglyceride components of the oil volatize and are stripped from the oil with the aid of injected steam. Properly deodorized oil will have a free fatty acid level of less than 0.05% and retain no oxidation products. The flavor of deodorized oil is relatively bland, with a slight nutty or buttery taste.

DEODORIZED OIL STORAGE

If not transferred directly from deodorization to a tank car or truck for shipment, deodorized oil is held in storage tanks prior to shipping. Effective deodorized oil storage requires the following to assure product quality:

1. nitrogen blanketing of the head space above the oil level
2. agitation to prevent separation of the triglyceride components
3. temperature control capability
4. sampling capability (Strahman sample valves are the industry standard)
5. tank washing and cleaning at the proper frequency to prevent the accumulation of "old" oil films on the interior tank surfaces.

Deodorized Oil Loading

The bulk loading methodology is a critical element in the delivery of acceptable quality oil to the end user. Tank cars or trucks must be clean and thoroughly inspected before oil loading begins. Hoses, nozzles, and pipelines must also be clean and properly protected.

Nitrogen sparging is recommended to minimize air exposure to the oil during loading. Sparging involves the introduction of nitrogen into the oil stream at a point before the loading nozzle (see Fig. 4.2). The subsequent release of nitrogen from the oil in the transport vessel provides a protective barrier against the air above the oil surface.

Bottom filling of tank cars or trucks using a loading nozzle extended through the top hatch is preferable, and a protective cover around the open top hatch during loading is essential.

Security seals properly attached to all access openings to the tank car or truck are mandatory after loading and sampling are completed. The security seals are protection against oil theft or sabotage.

Figure 4.2

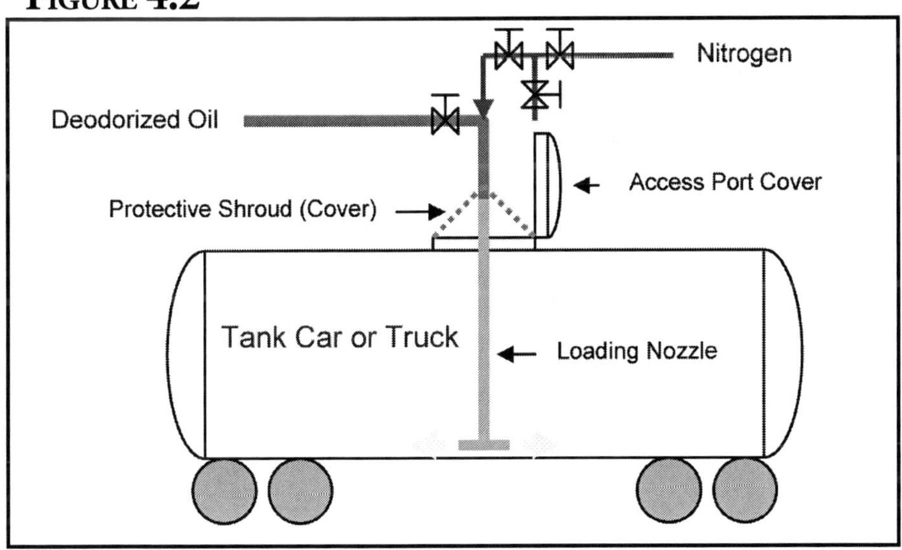

5 Unloading and Storage of Edible Oils

NORMAN J. SMALLWOOD

THE CORE TEAM (PARTNER)

When shipments arrive, the end user must take action to confirm that the oil quality is within specification and to protect the oil during unloading and storage. A reliable protocol must be established between the supplier and end user for verifying that each shipping vessel is used exclusively for transporting food products and is rigorously inspected before loading the product. Then, for each edible oil delivery, the responsibility is to thoroughly comply with this protocol by following the documentation and inspection procedures specified. If a transport vessel is not in compliance with the standards defined in the protocol, the shipment must be rejected.

The next step is to verify that every access port and nozzle into the transport vessel is security sealed to assure that no tampering has occurred. Seal numbers should be recorded and checked against those recorded on the shipping documents from the supplier. If an access port or nozzle is unsealed, the shipment must be rejected.

Quality evaluation of incoming edible oil shipments should be carried out promptly by properly trained and qualified personnel. The technicians assigned should understand the principles of proper sampling and be familiar with precautions that must be taken to protect the samples from contamination and abuse during the sampling process.

In taking the sample, the technician should clean off all extraneous material around the sampling port before opening the vessel. The sampling tools and containers must be clean and dry to avoid sample contamination. The temperature of the oil in the transport vessel should be determined in order to properly evaluate the sample. This is particularly important if the oil is partially or completely solid at ambient temperatures. If the product is partially solidified, the technician should arrange for low-pressure steam (< 20 psig) or hot water to be circulated in the coils of the transport vessel. High-pressure steam should never be used to heat an incoming oil shipment.

Using a zone sampling device, the technician should draw three samples from the transport vessel, one each from the top, middle, and bottom. A zone sampler is simply a small metal container with a lid that can be lowered to any level of the vessel. The technician lowers the sampler to the desired depth and opens the lid using a hand-held control, allowing the oil to enter the sampler. Any water present in the vessel will separate from the oil and settle to the bottom. If present, water and other sediment will be contained in the bottom sample.

Each of the three samples should be immediately placed in a sealable container, protected from light, and promptly taken to the quality control laboratory for analysis. Light exposure for only a short time can affect the flavor of the oil.

In the quality control laboratory, each sample should be visually examined to detect any anomaly. Careful visual examination of the bottom sample is especially important. Then, the three samples should be combined and subjected to the analyses for oil quality that are documented in the purchase specification. Most edible oils are delivered with a maximum peroxide valve (PV) specification of 1 mEq/kg (for tank trucks that contained the oil for less than 24 hours) or 2 mEq/kg (for tank cars), a free fatty acid level of 0.05% or less, a color appropriate to the oil, and a bland flavor typical of the oil. Oils that are solid at ambient temperatures can also be tested to determine that the melting point is appropriate for the oil.

Gas-liquid chromatography has become common in many quality control operations, and many laboratories routinely run a fatty acid distribution on methyl esters prepared from the incoming oil sample to ensure that the distribution is in the proper range for that oil.

If a significant quality deviation is found, a second sample should be taken and analyzed. If the analytical results of the second sample fail, the shipment should be rejected. Any further sampling and analysis of the oil would be counterproductive.

When it is found that the critical analytical results meet the purchase specifications, authorization can be given to unload the oil. Tank cars should not be unloaded until all the analyses are completed. In the case of tank trucks, where demurrage charges for unloading delay can be incurred, unloading after partial analysis is usually permissible if the partial results (free fatty acid, peroxide value, filter test, flavor, and refractive index) compare favorably with the supplier's certificate of analysis and meet product specifications.

Unloading on the basis of the supplier's certificate of analysis is not a prudent practice. The risk far outweighs the benefit.

It is important to record the results of the sample analysis for future reference. The statistical comparison of these records can be used to uncover problems and to help determine corrective action. Both the end user's and supplier's analytical capability can be tracked. The data will reflect the overall quality control competence of the laboratories involved. In addition, the information is useful in helping to optimize frying conditions. Correlations can be established over time between oil characteristics, fry life, and shelf life.

Unloading Operations

The unloading operations are critical to the quality attributes of edible oil. Lack of attention to detail in the unloading operations can cause irreversible damage to the oil. As noted, only low-pressure steam or hot water should be used to heat the oil in the transport vessel if necessary, and the oil should be heated to slightly above the melting point.

As shown in Fig. 5.1, the unloading system consists of an unloading hose, transfer line connecting the unloading hose to the suction of the unloading/transfer pump, transfer line from the pump discharge to an in-line filter, and a transfer line from the in-line filter discharge to the storage tanks.

Nitrogen should be used to purge air from the flow path before unloading is started. At the conclusion of the unloading operation, nitrogen should be used blow out the residual oil in the flow path.

The in-line filter is vital to remove the normal level of particulate matter inherent to every transporting and unloading system. A bag-type filter with a 5–10 micron filter element is most commonly used. The filter should be sized according to the unloading rate desired.

The unloading system should be designed and maintained to avoid air leakage into the product stream. Pump seals should be monitored for leakage, because

FIGURE 5.1. PROCESSED EDIBLE OIL HANDLING RECEIVING, UNLOADING, STORING, AND TRANSFERRING FOR USE

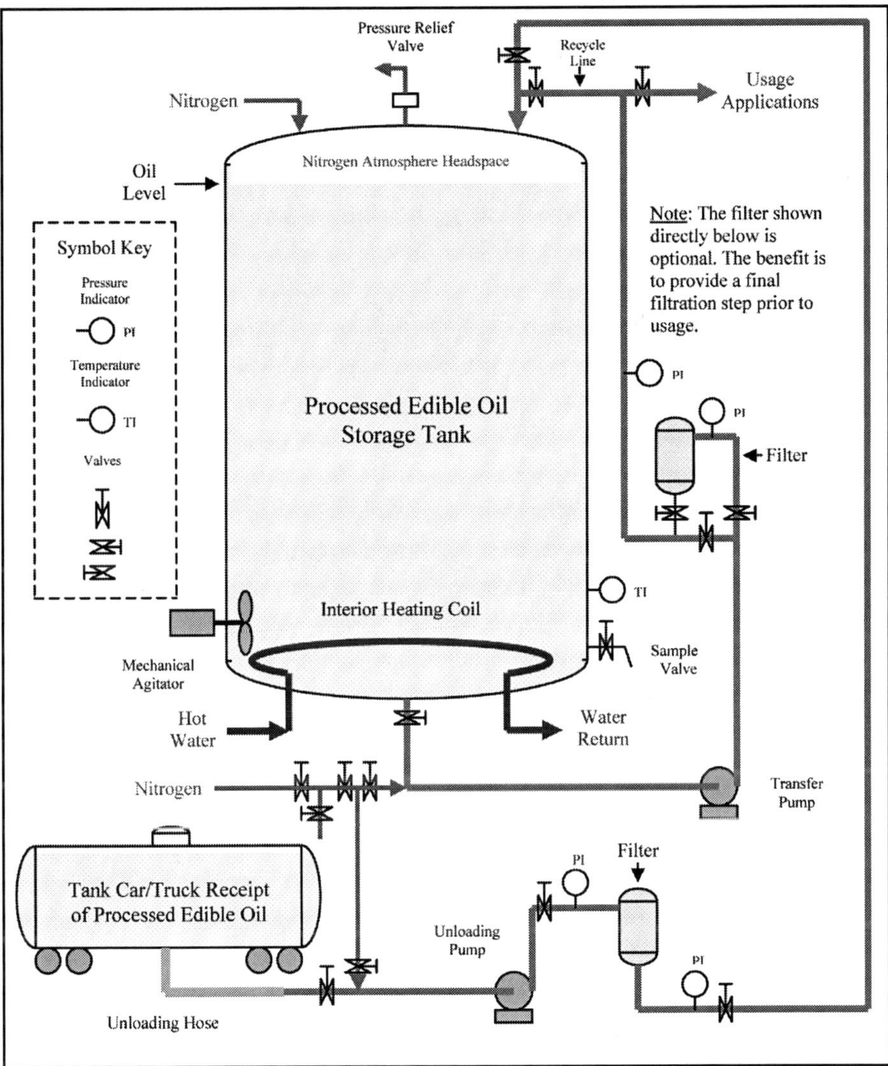

this is a point where air can enter the oil stream. The same applies to any connection or valve on the suction side of the pump. Replace the loading hose when it becomes excessively worn or cracked to the point that air leakage could occur.

Unloading hose ends must be equipped with protective caps to keep out foreign matter when not in use. Furthermore, the unloading hose should be secured when not in use to preclude being run over by a tank truck or tank car.

The operating sequence for unloading a tank car or truck of oil is as follows:

1. Confirm that the oil has been sampled, analyzed, and approved for unloading.
2. Identify the oil type and oil storage tank to be used.
3. Check the supplier's documentation to verify the quantity of oil in the shipment. Confirm that space is available in the storage tank to hold the oil.
4. Place a new element in the in-line unloading system filter at the proper frequency as determined by measuring the pressure drop across the filter.
5. Equalize pressure in the tank car or truck headspace to avoid catastrophic failure of the vessel due to a vacuum. This involves opening the top hatch of the tank car or truck to allow air entry into the void space as the oil is removed. A protective cover must be placed over the exposed port to prevent the entry of foreign matter.
6. Connect the unloading hose to the vessel outlet nozzle. When connecting the unloading hose to the tank car or truck outlet nozzle, the first step is to remove the cover from the nozzle. This should be done very slowly to avoid oil spillage if the outlet valve of the transport vessel is defective. If the valve has failed, the oil pressure will bear directly on the cover and the pressure can be detected before the cover is completely removed. To unload a tank car or truck with a defective outlet valve, leave the nozzle cover in place to prevent oil spillage. Then, extend the unloading line through the vessel top port and down to the bottom using a thoroughly cleaned extension hose or pipe.
7. Purge the unloading flow path with nitrogen if available to remove the air.
8. Set the unloading system valves to transfer the oil to the designated storage tank.
9. Start the unloading pump.
10. Confirm oil flow to the designated storage tank.
11. Check the pressure drop across the in-line filter several times during unloading.

12. If the pressure drop is excessive, shut off the pump and replace the filter element.
13. Confirm by visual inspection that the tank car or truck is completely unloaded.
14. Clear the unloading/transfer line system (flow path) of oil by introducing nitrogen or air pressure. Nitrogen is preferable, but air can be used if nitrogen is not available.
15. Disconnect the unloading hose and cap the open end.
16. Secure the unloading hose to prevent damage in the traffic area.
17. Measure the quantity of oil in the storage tank and compare the volume gain to the quantity of oil contained in the transport vessel.
18. Obtain and analyze a representative sample of oil from the storage tank. Compare the results from the tank sample to the results from the pre-unloading sample. Be aware that a representative sample of oil cannot be obtained from a storage tank unless mechanical agitation of the oil is provided

OIL STORAGE TANKS

The receiving tank for the oil storage must be clean and dry. If nitrogen blanketing is not available, the inlet (filling) pipe should extend to the bottom of the tank to minimize oil aeration and oxidative degradation.

Static storage of oil results in stratification based on specific gravity differences in the oil components. Water and any particulate matter settle on the tank bottom. The consequence of static storage is that each transfer of oil to the usage point will vary. The quality differences in each oil lot transferred can be sufficiently different to negatively impact performance.

Where static oil storage is used, it is not an uncommon practice to circulate the oil from tank outlet to tank inlet for a period of time for the purpose of achieving homogeneity before transfer to usage. Such a practice is ineffective because the specific gravity stratification rate is greater than the pumping rate for the capacity of most storage tanks.

Equipping oil storage tanks with mechanical agitation is highly recommended to assure oil homogeneity. Another benefit of mechanical agitation is to avoid localized overheating if it is necessary to store the oil above ambient temperature.

To achieve oil homogeneity with a single side-entering mechanical agitator (the most economical solution), the tank height cannot exceed two times the tank diameter. If this principle is violated, there will be stagnant zones in the tank.

If it is necessary to maintain the storage tank at an elevated temperature, only low-pressure steam or hot water should be used to heat the oil. And, the temperature should be maintained at the lowest practical point. Appropriate automatic temperature control should be utilized to avoid overheating the oil in storage.

Storage tanks should be constructed for complete emptying by means of a dish, cone, or sloped bottom. Flat-bottom tanks are unacceptable for finished oil storage because of the inability to completely empty the tank between shipment receipts.

Oil storage tanks should be inspected quarterly and thoroughly washed at least once per year. For each cycle of use (tank filling and emptying), a film of oil is deposited on the interior tank surface, similar to a new coat of paint. When the films of oil accumulate to the point that erosion occurs, the old oil films sloughing off the tank walls and become part of the product stream. The consequence is oil quality degradation. Thus, failure to maintain clean tanks will result in poor oil and product quality.

Oil Transfer to Fryers and Cookers

The critical consideration in transferring oil from storage tanks to fryers is to avoid air contact with the hot oil. Air contact can be eliminated by nitrogen purging the flow path before transferring the oil. The air exposure problem can be minimized by nitrogen sparging (injecting nitrogen into the oil stream on the discharge side of the transfer pump) during transfer.

For the case of no nitrogen availability, the air can be cleared from the pipeline to the cooker or fryer by circulating oil in a closed loop (storage tank to a point near the cooker or fryer inlet and back to the storage tank). After brief circulation, the oil flow can be diverted to the fryer or cooker. This will confine the air exposure to the oil at storage temperature instead of the much higher fryer or cooker temperature.

The oxidation rate of edible oil is accelerated when the oil temperature rises. When oil from a fryer is transferred to a storage tank to allow for fryer cleaning or servicing, exposure of high-temperature oil to air is damaging. Air exposure is also possible when oil is circulated from a fryer through an empty filter to remove food particles. In both cases, nitrogen purging and blanketing should be used to minimize oxidation.

Use of nitrogen for protecting oil from oxidation is a potential personnel safety hazard if proper precautions are not followed. Tanks or vessels under nitrogen blanketing or sparging cannot be entered, and appropriate safeguards and warning signs must be used. Where nitrogen is introduced into confined building space, sufficient ventilation must be provided to maintain a normal oxygen level in the atmosphere where people are present.

6 Factors Contributing to Oil Deterioration

THOMAS G. CROSBY

FRITO-LAY, INC.

Preventing deterioration of edible oil during frying is extremely important in producing quality snack food products. The three major ways in which oils deteriorate—hydrolysis, oxidation, and polymerization—are covered in the following sections.

HYDROLYSIS

This reaction is just what the word implies—chemical bonds that hold the triglycerides together are broken by water, creating free fatty acids, monoglycerides, diglycerides, and even glycerine if the hydrolysis is extensive.

In frying situations, hydrolysis reactions are often promoted by surface active agents and excessive moisture in the products being fried. Caustic materials such as acids or bases are excellent catalysts of the hydrolysis reactions. It is extremely important to remove all traces of such materials from the fryers after cleaning.

Metal ions are also catalysts for hydrolysis, and can react with free fatty acids to form soaps. Using one of several available methods to monitor the free fatty acid level of the oil is advisable, although a problem may exist even if the free fatty acid is within an acceptable range.

Frying systems with high oil turnover rates usually maintain acceptable oil quality, and released steam also tends to strip free fatty acids from the oil. It is important that the fryer hood and vent systems be designed so that "dripback" does not occur. If moisture and free fatty acids condense above the fryer and drop back into the oil, hydrolysis and flavor deterioration are likely to occur.

OXIDATION

Edible oils react with oxygen in the air at all temperatures, but these reactions increase significantly as the temperature increases. The primary reactions are the formation of hydroperoxides adjacent to the double bonds in unsaturated fatty acid molecules.

As the number of double bonds in the oil is increased, the rate of oxidation also increases. As a rule, if the oxidation rate of oleic acid is taken as one, the oxidation rate of linoleic acid is 10, and the oxidation rate of linolenic acid is 100 (based on oxygen uptake). This gives a clue as to why various oils have different oxidative stabilities.

If the only result of oxidation was the formation of hydroperoxides, the problem would not be extreme. However, these hydroperoxides undergo further reactions that generate more free radicals, which lead to chain reactions that produce even more of these byproducts. Hydroperoxides also break down into aldehydes, ketones, and other compounds that cause flavor problems.

Oxidation is also enhanced by heavy metal ions, such as those found in copper and iron. Copper is a potent oxidation catalyst and must be avoided at all costs. No copper fittings or brazings can be permitted in frying equipment because the fry life of the oil will be diminished significantly. Operators must be alert and check any new equipment or modifications to existing equipment to ensure that oils are not exposed to copper or materials that contain copper.

Polymerization

Another reaction of unsaturated fatty acid molecules is polymerization. In practice, it leads to gummy deposits on frying equipment and a greasy appearance in fried products. In normal frying equipment at normal temperatures, the major type of polymerization that occurs is called "oxidative polymerization."

The free radicals formed when hydroperoxides decompose at high temperatures combine, doubling the weight of the triglyceride precursors. As such reactions continue, the molecular weight of the polymers increases and these products are no longer soluble in the oil. When the polymers are deposited on the fryer wall, they react readily with oxygen in the air and with more fatty acid molecules in a chain reaction.

The other type of polymerization is "thermal polymerization." This occurs primarily at "hot spots" in frying equipment, caused by localized overheating.

It is very important to remove these polymers from the equipment if they occur, but mechanical scrubbing alone is not enough. Highly caustic solutions must be employed to hydrolyze the polymers so they can be flushed out of the frying equipment. An acid rinse to neutralize the caustic residue followed by adequate clear water rinses must then be used so that no trace of acid or caustic remains in the fryer.

In most high-turnover systems using moderately stable oils, polymer formation should not be a problem. In instances where polyunsaturated fatty acids are used or where temperatures and/or oxygen exposure are factors, operators should be alert to the possibility of polymer formation and the deterioration of product quality that can result.

Use of Edible Oil in the Cooking Process

Don Banks

Edible Oil Technologies

Edible oils comprise between 15 – 44% of a snack food product, depending on the product and preparation process involved. For this reason the oil has a significant impact on the flavor, eating quality, shelf life, and other attributes of finished snacks.

Oil Flavors

Before processing, edible oils have relatively strong flavors with characteristic aromas primarily derived from minor components unique to the origin of each type of oil. Modern refinery processing is designed remove all flavor and aroma from oils to produce completely bland products. With appropriate packaging or bulk storage, oil will remain bland for an extended period of time. However, as oils are subjected to oxidation the characteristic flavor derived from minor components can redevelop and be accompanied by additional flavor components derived from the oxidation of fatty acids. The extent to which flavor develops from both sources is significantly influenced by the quality of the oil before processing and by how carefully quality control is maintained during refining, storage and shipment of the finished oil.

The flavor character of some oils such as peanut, cottonseed, and to some extent corn is commonly recognizable and, in the early stages of oxidation, can enhance the overall flavor of some products. The characteristic flavor of other oils such as sunflower, palm, and safflower are usually more subtle. Canola and soybean are different in that the green, grassy, and beany flavor commonly associated with these oils occur in the early stages of oxidation and are primarily derived from the oxidation of linolenic acid. However, new varieties of both of these oils have recently been developed that contain 3% or less linolenic and do not develop objectionable flavor during early oxidation.

Effects of Frying

When oils are partially hydrogenated, they develop a characteristic "hydrogenation" flavor. Depending on the extent of hydrogenation, the characteristic flavor that the oil would otherwise develop during aging is significantly reduced or eliminated.

When fresh oil is first placed in a fryer it is very bland. During initial frying, flavor is developed from starch, amino acids, and other components in the product being fried to generate flavor characteristic of the product. That is why potato chips basically taste like potato chips and corn chips taste like corn chips even when different types of frying oils are used. Oil that has been properly processed does not provide a significant flavor contribution at this point; rather it slightly reduces flavor intensity of the fried product by dilution. During a conditioning period, product flavor will accumulate in the oil to the point that it tastes just like the finished product being fried.

During continuous frying, the free fatty acid (FFA) will increase from an initial value (i.e. ~0.03%) to an equilibrium value, usually between 0.12 and 0.40%, depending on the operating parameters and frying system being used. The FFA does not directly contribute to product flavor but it is an indicator that the oil is being subjected to hydrolytic and oxidative changes and that aldehydes, ketones, short-chain fatty acids, and other compounds are being formed. As these breakdown components build up in the oil, they begin to contribute to the flavor profile of the finished product.

With modern continuous fryers that are operated with rapid "oil turnover" (oil turnover = the time required for the addition of makeup oil to equal the volume of oil used to fill the fryer) and with good oil filtration, the FFA will equilibrate around 0.10–0.14%. Under these conditions, the used oil in the fryer undergoes very limited change, and analytically remains quite similar to fresh oil. In this case, the oil primarily acts to concentrate and carry the characteristic flavor of the product and provides only subtle contribution to the flavor profile of the finished product.

With older fryers that generate more oil stress and have limited means for crumb control, the FFA commonly equilibrates in the range of 0.25–0.40%. Under these conditions, the frying oil can generate increased levels of degradation products. In such case, flavor differences attributable to the type of frying oil being used can readily contribute, for better or worse, to the flavor profile of the fried product.

Unlike continuous fryers, batch fryers usually do not have a rate of oil turnover sufficient to maintain equilibrium conditions, and the oil degrades over time. As the oil degrades, the flavor contribution from minor components in the oil becomes stronger and then flavor notes from the oxidation of fatty acids begin to build. Timely replacement of the oil will be needed or objectionable off-flavors, and eventually rancidity, will occur.

Oils that are liquid at room temperature support rapid flavor release and maximum flavor intensity during mastication of snacks, as compared to oils with higher melting points. Oils that are solid at room temperature but melt below body temperature will generally have greater oxidative stability than common liquid oils, but flavor release is slower and intensity is moderately reduced. High stability oils with melting points above body temperature support a "dry" nature (non-oily appearance and dry eating quality), but they can delay flavor release and mute flavor intensity. In some applications, higher melting oils also exhibit a waxy mouth feel.

Slurry and spray oils are used topically and should be sensory tested to ensure they enhance the finished snack foods and do not detract from them. Since these oils are applied to the surface of products, any flavor or off-odor picked up during handling or storage will be readily apparent. The temperature of the spray oil should be maintained as low as possible to ensure good coverage with minimum oxidation.

When evaluating an alternative frying oil for an established product, careful testing needs to be conducted to obtain meaningful and reliable results. First, the oil must be analyzed to be certain that it meets the quality standards specified for the oil. Then the oil needs to be conditioned by continuous usage until the FFA reaches a nominal, steady state value. Product can then be sampled for evaluation initially and under simulated storage and display conditions throughout the shelf life of the product. Ideally comparative testing should be conducted with control product produced under identical conditions.

OTHER EFFECTS ON OIL

Heating, condensate, crumb load, aeration, and other factors will also have a significant effect on the quality of oil and subsequent flavor of the finished snack food.

Oil has to be heated for frying but heating should not be initiated any sooner than necessary to start production on schedule. The timing should be controlled to start production as soon as the oil first reaches frying temperature. Any delay in starting frying will subject the oil to unnecessary oxidation and degradation. The discharge of steam and the addition of makeup oil during frying help preserve oil quality. Any time frying is interrupted the benefits are lost and the oil is subjected to increased stress. To minimize the degradation, whenever there is an interruption in frying that will last more than a few minutes, heating should be discontinued and steps taken to minimize oil aeration, consistent with operating requirements specified by the fryer manufacturer.

Aeration of the oil at frying temperature should be minimized because oxidation occurs at a much faster rate at these temperatures. Aeration is a primary source of oxidized flavors in the finished product. The manifold that returns heated oil to the fryer should be adjusted for optimum flow with minimum aeration of the oil.

Condensate must be carefully controlled to prevent drip-back from contaminating the oil. The draft under the hood of modern fryers ensures the removal of most condensate. However, if the condensate pan is not positioned properly or if the drainage system is not working properly, condensate can drip back into the fryer, giving the oil a dark color and a bitter, acrid flavor.

Crumb load will influence the color, flavor, and FFA of the oil and excessive accumulations can impede fryer operation. The accumulation of crumbs should be minimized to ensure that oil quality is maintained. Excessive buildup can interfere with heat transfer, causing blockage in heat exchanger systems or causing hotspots in direct-fired fryers, and has been known to cause fryer fires.

A variety of systems are available to control crumb accumulations. Sidestream filtration in a closed loop system has been commonly used over the years. A variety of newer systems including rotating drum, centrifugal separation, continuous belt, continuous paper, and others can be used for full-flow filtration to provide increased crumb control.

Other factors also affect the quality of oil during frying. For example, the seasoner should be located far enough from the fryer hood draft to prevent seasoning from being drawn into the fryer. The manifold that returns heated oil to the fryer should be adjusted for optimum flow with minimum aeration of the oil.

To minimize oil oxidation that occurs at the end of production, water-cooled heat exchangers can be used to rapidly cool the oil. (Heat exchanger systems are particularly beneficial for use with large volume fryers.) As soon as the oil has been sufficiently cooled it should be filtered and transferred to storage. Nitrogen can be used to protect the used oil from oxidation during storage between periods of production.

An especially effective technique for using nitrogen to protect oil during storage is to slowly introduce nitrogen to the bottom of the tank and vent it at the top of the tank. The nitrogen will strip oxygen and volatile oxidation components as it rises through the oil, enhancing oxidative stability. When using this technique, it is essential that the nitrogen be ultra pure (i.e., 99.99%).

Oil should be filtered again when it is transferred from storage back to the fryer. Filtration is recommended because the additional time and cooling during storage supports the removal of polymers that formed during frying. These polymers are a source of varnish-like flavors.

The oil content of the finished product depends on numerous variables including: type of oil, melting point, orientation of the product on the takeout conveyor, takeout conveyor speed (drain time), etc.

The remainder of this chapter reviews some of the oils commonly used for each type of snack product, the oil content ranges, cooking oil temperatures, and factors influencing oil content (Fig. 7.1 – 7.8).

Figure 7.1. Potato Chips

<table>
<tr><td>

TYPES:

    Plain
    Ridged
    Baked
    Fabricated

OILS USED:

    Cottonseed
    Corn
    Peanut
    Canola
    Soybean
    Partially Hydrogenated Soy
    Others: Palm, Palm Olein, Sunflower

</td><td>

OIL CONTENT:

Fried	Range	28 – 44%
	Average	32 – 36%
Baked	Range	4 – 9%
	Average	5 – 7%

OIL TEMPERATURE RANGE: 350 – 370°F

OIL ABSORPTION INFLUENCED BY:

    Potato variety
    Specific Gravity of Potato
    Condition of Potato
    Slicer Operation
    Slice Washer
    Oil Temperature
    Frying Time
    Takeout Belt

</td></tr>
</table>

Figure 7.2. Kettle-Style Chips

<table>
<tr><td>

TYPES: Kettle-Style

OILS USED:

    Cottonseed
    Peanut
    Partially Hydrogenated Soy
    Others: Palm, Palm Olein, Sunflower

OIL CONTENT: Range   24 – 44%
               Average 28 – 38%

</td><td>

OIL TEMPERATURE RANGE: 270 – 340°F

OIL ABSORPTION INFLUENCED BY:

    Potato variety
    Specific Gravity of Potato
    Condition of Potato
    Slicer Operation
    Oil Temperature
    Frying Time
    Takeout Belt

</td></tr>
</table>

Figure 7.3. Corn Chips

<table>
<tr><td>TYPES:</td><td>Cooked Corn
Masa Mix</td><td>OIL CONTENT:</td><td>Range
Average</td><td>26 – 36%
28 – 32%</td></tr>
<tr><td>OILS USED:</td><td>Corn
Canola
Sunflower
Soybean
Partially Hydrogenated Soy
Others: Palm Olein</td><td>OIL TEMPERATURE RANGE:</td><td></td><td>375 – 415°F</td></tr>
</table>

OIL ABSORPTION INFLUENCED BY:

Cooked Corn	**Masa Mix**
Corn Cooking Conditions	Masa Formulations
Steep Time Conditions	Make-up Procedures
Grinding	> Mixing
Moisture Content	> Moisture
Extruder Operation	> Temperature
Oil Temperature	> Floor Time
Frying Time	Extruder Operation
Takeout Belt	Oil Temperature
	Frying Time
	Takeout Belt

FIGURE 7.4. TORTILLA CHIPS

TYPES: Cooked Corn (yellow, blue, white)
          Masa Mix

OIL CONTENT: Range   22 – 36%
                  Average   24 – 32%

OILS USED: Corn
            Canola
            Soybean
            Sunflower
            Soybean and
            Partially Hydrogenated Soy Blend
            Partially Hydrogenated Winterized Soy
            Partially Hydrogenated Soy
            Others: Palm, Palm Olein

OIL TEMPERATURE RANGE: 340 – 365°F

OIL ABSORPTION INFLUENCED BY:

Cooked Corn
- Corn Cooking Conditions
- Steep Time Conditions
- Grinding
- Sheeting
- Oven Conditions
- Equilibration (Tempering)
- Oil Temperature
- Frying Time
- Takeout Belt
- Spray Oil Rate

Masa Mix
- Masa Formulations
- Makeup Procedures
  - > Mixing
  - > Moisture
  - > Temperature
  - > Floor Time
- Sheeting
- Oven Conditions
- Equilibrium (Tempering)
- Oil Temperature
- Frying Time
- Takeout Belt
- Spray Oil Rate

FIGURE 7.5. EXTRUDED SNACKS

TYPES:	Fried	OIL CONTENT:	Range	21 – 44%
	Baked		Average	32 – 38%
	Popped Pellets			

OILS USED:
Corn
Canola
Soy
Sunflower
Peanut
Coconut
Partially Hydrogenated Soy
Partially Hydrogenated Soy/Cottonseed
Partially hydrogenated Cottonseed
Others: Palm, Palm Olein

OIL TEMPERATURE RANGE:
Frying   340 – 370°F
Spray Oil   90 – 130°F

OIL ABSORPTION INFLUENCED BY:

Fried	**Baked**
Corn Meal	Corn Meal
Extruder Operation	Extruder Operation
Oil temperature	Oven Conditions
Frying Time	Slurry Makeup
Slurry Makeup	Coating Tumbler Operation
Coating Tumbler Operation	

FIGURE 7.6. POPCORN

TYPES:	Hot Air Popped	OIL CONTENT:		
	Hot Oil Popped		Range	15 – 35%
			Average	24 – 30%

OILS USED:
Coconut
Canola
Partially Hydrogenated Soy
Partially Hydrogenated Cottonseed
Others: Palm

OIL TEMPERATURE RANGE:
Hot Oil Popped   395 – 425°F
Spray Oil   90 – 130°F

OIL ABSORPTION INFLUENCED BY:

Hot Air Popped	**Hot Oil Popped**
Spray Oil Rates	Oil/Corn Ratio
Slurry Makeup	Temperature of Popper
Coating Tumbler Operation	Proper Loading
	Expansion of Kernels
	Slurry Makeup
	Coating Tumbler Operation

FIGURE 7.7. NUT ROASTING

TYPES:	Peanuts
	Other Nuts
OILS USED:	Peanut
	Coconut
	Partially Hydrogenated Cottonseed
	Partially Hydrogenated Soy
	Others
OIL CONTENT:	2% Oil Uptake

OIL TEMPERATURE RANGE: Range   300 – 360°F

                                         Average   310 – 325°F

OIL ABSORPTION INFLUENCED BY:

       Roaster Operation
- Oil temperature
- Frying Time
- Roaster Load

       Spray (Dressing) Oil

FIGURE 7.8. PORK SKINS

TYPES:	Pellets
OILS USED:	Lard
	Lard and Hydrogenated Vegetable Oil

OIL CONTENT: Range   26 – 40%

                         Average   32 – 38%

OIL TEMPERATURE RANGE:

              Range: 375 – 410°F

              Average   385 – 395°F

OIL ABSORPTION INFLUENCED BY:

       Pellets
       Fryer Load
       Oil Temperature
       Frying Time

8 Cleaning Edible Oil Processing Equipment

RICK DELLA PORTA

FRITO-LAY, INC.

The importance of maintaining clean and sanitary food processing equipment cannot be overemphasized. This is especially true in oil processing equipment, where unseen contaminants can quickly ruin both edible oil and food processed in it. Regularly scheduled sanitation procedures are essential to producing high-quality snack food products.

Proper sanitation procedures will also ensure that a manufacturing plant meets applicable regulations of the Food and Drug Administration, Environmental Protection Agency, American Institute of Baking, United States Department of Agriculture, and any state or local regulatory agencies.

Manpower and utility costs comprise about 94% of the sanitation dollar—only about 6% of the cost goes to cleaning chemicals. For this reason, it is important to make sanitation an integral part of the operation.

An effective sanitation program begins with cleaning. Proper cleaning of metal surfaces will reduce the number of microorganisms present and create an environment where they cannot thrive. Cleaning will also eliminate other contaminants, increase the life of the equipment, and add to the aesthetics of the operation.

Simply defined, cleaning is a process which will remove soil and prevent accumulation of food residues which may decompose or support the growth of disease or nuisance-causing organisms and/or the production of toxins.

EVALUATING THE WATER SUPPLY

Water is the key ingredient in all food processing plant cleaners. It must be free from disease-producing organisms, toxic metal ions, and objectionable odors and tastes. No food processing plant has an ideal water supply; therefore, cleaning compounds must be tailored to fit the water supply and type of operation.

For example, suspended matter must be kept to a minimum to avoid deposits on clean equipment surfaces. Concentrations of soluble iron and manganese salts above 0.3 ppm will cause colored deposits on equipment surfaces. Suspended matter and soluble iron and manganese can be removed only by treatment.

Water hardness caused by salts of calcium and magnesium also form surface deposits and diminishes the effectiveness of cleaners. Carbonate hardness (formerly called temporary hardness) is due to calcium and magnesium carbonates and bicarbonates and can be removed by heating. Noncarbonated hardness (formerly called permanent hardness) is due to calcium sulfates, calcium chloride, magnesium sulfate, and magnesium chloride, which cannot be removed by heating.

THE CLEANING PROCESS

Table 8.1 describes the common cleaning functions achieved during the cleaning process. Specific types of cleaning compounds will be required to achieve the cleaning functions required for a particular cleaning job. However, before reviewing these requirements it is helpful to understand the four steps that apply to most cleaning situations.

1. Bring the detergent solution into contact with the soil to be removed by sufficient wetting and penetration of the detergent.
2. Displace the solid and liquid soils from the surface to be cleaned by saponifying the fat, peptizing the proteins, and dissolving the minerals.
3. Disperse the soil in the solution by dispersion or emulsification.
4. Prevent return of the dispersed soil to the clean surface by sufficient rinsing.

Cleaning compounds and cleaning methods will vary depending on the type of soil on the surface to be cleaned. Table 8.2 shows some different food constituents and their characteristics that must be considered before cleaning begins. Stainless steel, aluminum, plastic, and painted surfaces all require different compounds.

To understand the functions performed by different cleaning agents, see Table 8.3.

TABLE 8.1. CLEANING FUNCTIONS

Some of the common functions achieved during the cleaning process include:

Chelation / Sequestration	The action of a chemical compound attaching itself to the water hardness particles, inactivating them and preventing them from combining with other materials in the water and precipitating out.
Dispersion	The action in which groups or clumps of particles are broken up into individual particles and spread out or suspended in the solution.
Dissolving	The reaction that produces water-soluble materials from insoluble soil.
Emulsification	The dispersion or suspension of fine particles or globules of one or more liquids in another liquid.
Peptization	The softening or liquefaction of one substance by trace quantities of another, analogous to the digestion of proteins.
Precipitation	The act of softening water by converting hardness minerals into an insoluble form.
Rinsability	The ability to freely separate soil from the surface when fresh water is flushed over it.
Saponification	Action of alkali on fats resulting in the formation of soap.
Suspension	The action in which insoluble particles are held in solution.
Wetting	An action to lower the surface tension of the water medium to increase its ability to penetrate soil.

ACID-TYPE CLEANERS

Acid-type cleaners are widely used for cleaning in the food industry. Mineral deposits on equipment are nearly impossible to remove with alkaline cleaners and these cleaners may contribute to mineral deposits. Consequently, acid-type cleaners are used to remove these deposits.

A wide variety of acid-type cleaners are available. They are blends of organic acids, inorganic acids, or acid salts, usually including wetting agents. To be

TABLE 8.2. CLEANING FOOD CONSTITUENTS FROM EQUIPMENT

Component on Surface	Solubility Characteristics	Ease of Removal	Changes Induced by Heating Soiled Surface
Sugar	Water soluble	Easy	Carmelization, more difficult to clean
Fat	Water insoluble, Alkali soluble	Difficult	Polymerization, more difficult to clean
Protein	Water insoluble, Alkali soluble, Slightly soluble in acid	Very difficult	Denaturation, much more difficult to clean
Salts: – Monovalent	Water soluble, Acid soluble	Easy	None
Salts: – Polyvalent (i.e. $CaPO_4$)	Water insoluble, Acid soluble	Difficult	Interactions with other constituents, more difficult to clean

The differing characteristics of food constituents affects the choice of cleaning compounds used for equipment cleaning. This chart shows how some common food constituents react to the cleaning process.

effective, an acid-type detergent should produce a pH level of 2.5 or lower in the final use solution. It should work well on hard as well as soft water and show minimal corrosion on metals.

The composition and concentration of the cleaning compound will depend on the nature and amount of soil on the surface and the type of surface.

The choice of solid or liquid cleaning compound influences the convenience and cost of cleaning operations. Liquid materials are frequently more hazardous to handle, but allow better concentration control through automatic feed devices. Powdered materials are more frequently overused and should be preweighed into standard amounts to improve efficiency and reduce waste.

If the water is heavily loaded with scale-forming materials such as calcium,

TABLE 8.3. FUNCTION OF CLEANING COMPOUNDS

Class of Compounds	Major Functions
Basic Alkalies	Soil displacement by emulsifying, saponifying, and peptizing
Acids	Mineral deposit control and removal; water softening
Surfactants	Wetting and penetrating soils; dispersion of soils and prevention of soil redeposits
Complex phosphates	Soil displacement by emulsifying and peptizing, dispersion of soil; water softening, prevention of soil deposits
Chelating Compounds	Water softening; mineral deposit control. Soil displacement by peptizing; prevention of redeposits
Oxidizing agents	Improves peptizing action of alkalies and provides bleaching and dissolving properties
Solvent	Degreasing action

This chart shows seven basic classes of cleaning compounds and their functions.

magnesium, iron, or sulfate, the cleaning compound must be adjusted to eliminate the depositing minerals, or the water must be treated to reduce the mineral content.

Efficiency of post-cleaning operations is also directly related to water quality. Mineral salts in rinse water will precipitate more readily from alkaline solutions than from acid. Conditioning the rinse water to a pH level of 6.5 or less will minimize mineral deposits on equipment surfaces.

FACTORS AFFECTING CLEANING

Whether cleaning by hand or using circulation cleaning, several factors affect the efficiency of cleaning operations:

1. *The Cleaner.* Variables in the type of cleaner used have already been discussed.
2. *Temperature.* Increasing temperatures decrease the strength of the bond between soil and surface, decrease viscosity, increase solubility of soluble materials, and increase chemical reaction rates.
3. *Velocity or Force.* In hand cleaning, physical pressure improves removal of soil from surfaces. In circulation cleaning, increased turbulence from faster fluid flow helps remove soil. Fluid should flow at a rate of 5 feet per second to ensure adequate turbulence.
4. *Time.* If all factors are constant, cleaning efficiency is improved by utilizing longer time periods.
5. *Concentration of Cleaner.* Increased concentrations of cleaners speed up chemical reactions. Among the variables listed here, this is the least effective in improving cleaning efficiency.

THE CLEANING CYCLE

The ideal cleaning cycle consists of the following steps:

1. pre-rinse
2. application of detergent solution
3. rinse
4. periodic acid rinse or cleaning
5. post-rinse
6. sanitizing rinse
7. final potable rinse, if required.

In some cases, acid cleaning must be flowed by alkaline detergent cleaning.

These steps are essential in all cleaning procedures, regardless of the method used. Pre-rinsing is important to minimize the soil load in the cleaning system and can effectively remove upto 90% of the soluble material. The cleaning operation loosens and removes the soil, and a post-rinse prevents redepositing of the soil on the clean surface.

Depending on the type of soil to be removed and method of chemical application, temperature selection for the pre-rinse, detergent solution, and post-rinse is very important. Usually tempered water (90–120°F) is used for both pre- and post-rinsing. The temperature of the chemical detergent solution will depend on the type of cleaning to be done. Consult your cleaning materials supplier for advice.

CLEANING METHODS

Large food particles should be removed before cleaning by flushing the equipment with cold or warm water under moderate pressure. Hot water or steam should not be used because it may make cleaning more difficult.

Several methods of applying cleaning compounds can be used. They include:

1. *Manual.* Usually a brush or abrasive tool is used to agitate the soil. This method provides excellent chemical contact.
2. *Soaking.* Small equipment, fittings, or valves may be immersed in cleaning solutions, and larger vessels, such as vats and tanks can be partially filled with a predissolved cleaning solution. The cleaning solution should be hot (125°F) and the equipment permitted to soak for 15–30 minutes before manual or mechanical scrubbing.
3. *Spray methods.* Cleaning solutions may be sprayed on equipment surfaces using either hot water or steam.
4. *Clean-in-place (CIP) System.* This method is an automated cleaning system generally used with permanent welded pipeline systems. Fluid turbulence in the pipeline is the major source of energy used for soil removal.
5. *Clean-out-of-place System.* Many small parts can be cleaned most effectively in a recirculating parts washer. A recirculating pump and distribution headers agitate the cleaning solution in a sanitary tank. In some cases, the parts washer may also serve as the recirculating unit for clean-in-place operations.
6. *Foaming.* A concentrated blend of surfactants are added to highly concentrated solutions of either alkaline or acid cleaners. This combination produces a stable, copious foam when applied with a foam generator. The foam clings to the surface to be cleaned, increases contact time of the liquid with the soil, and prevents rapid drying and runoff of the liquid cleaner, thereby improving cleaning.
7. *Gelling.* A concentrated powdered gelling agent is dissolved in

hot water to form a viscous gel. The desired cleaning product is dissolved in the hot gel and the resulting gelled acid or alkaline detergent is sprayed on the surface to be cleaned. The gelled cleaner will hold a thin film on the surface for 30 minutes or longer to attack the soil. Soil and gel are removed with a pressure warm-water rinse.

8. *High-pressure cleaning.* Hydraulic cleaning systems are frequently used to clean the exterior parts of equipment, floors, and some building surfaces. The cleaning compound is delivered through a high-pressure spray nozzle. Steam injection systems and pressure-fed tanks generally operate with nozzle pressures of 60–175 pounds-per-square-inch (psi). Air- and motor-driven high pressure pumps may develop nozzle pressures from 300–1,200 psi. Cleaning effectiveness depends on the force of the cleaning solution against the surface, which is controlled by the nozzle design.

SANITIZING

Sanitizing procedures are necessary to destroy any disease-causing organisms that may be present on equipment to prevent them from reaching the consumers. The procedures also prevent food spoilage. Even so-called harmless microbes can become a nuisance under certain conditions.

The following terms define some of the materials and procedures used in the sanitation process:

- *Sterilizer.* An agent that will destroy or eliminate all forms of life, including all forms of vegetative bacteria, bacterial spores, fungi, and viruses.
- *Disinfectant.* An agent that will kill 100% of most infectious bacteria, although not necessarily capable of killing bacterial spores.
- *Sanitizer.* A substance that reduces the microbial contaminants to safe levels as determined by public health requirements.
- *Bacteriostat.* An agent that inhibits or prevents the reproduction of bacteria.
- *Sanitization.* Application of any effective method or substance to a clean surface to destroy pathogens and other organisms. Such treatment shall not adversely affect the equipment, the product,

or the health of the consumer, and shall be acceptable to health authorities.
- *Cleaner-sanitizer.* A product that possesses the properties of both a cleaner and sanitizer.

Sanitizers can act either through heat (steam, hot water, or hot air) or chemicals. Table 8.4 shows four commonly used chemical sanitizers and lists the applicable levels, advantages, and disadvantages, of each.

REGULATORY CONSIDERATIONS

Compounds intended for use as antimicrobial agents marketed in interstate commerce must be properly labeled and registered by the Environmental Protection Agency (EPA) in accordance with the provisions of the Federal Insecticide, Fungicide, and Rodenticide Act as amended in 2004. Such regulations require manufacturers of such products to prove they are safe and effective for the intended use.

Before using sanitizers on surfaces always requiring a rinse, food products and packaging materials must be removed from the room or carefully protected. After using these compounds, surfaces must be thoroughly rinsed with potable water before food processing operations are resumed. The compounds must always be diluted and used according to the applicable directions on the EPA-registered label.

When sanitizers are used on surfaces *not* always requiring a rinse, food products and packaging materials must again be either removed from the room or carefully protected. However, a potable water rinse is not required following use of these compounds on previously cleaned hard surfaces provided that the surfaces are adequately drained before contact with the food. This ensures that little or no residue remains which can adulterate or have deleterious effect on edible products. A potable water rinse is required following use of these compounds under conditions other than those stated above. The compounds must always be diluted and used according to the applicable directions on the EPA-registered label.

A description of sanitizing solutions that may be used on hard, non-porous food-contact surfaces such as equipment and utensils can be found in "Guidelines for Obtaining Authorization of Compounds to be used in Meat and Poultry Plants," USDA Handbook No. 562 and in "Food Services Sanitation Manual," Section 178.1010, pages 83–87.

Proper cleaning of all processing equipment will help ensure the production of high-quality snack products with good shelf-life. Training of personnel in cleaning techniques and procedures is essential, and the cleaning should be made part of the regular production process, with specific time allotted for cleaning tasks.

Manufacturers that integrate good cleaning practices into their operations will find that the benefits far outweigh the costs.

FRYER BOIL-OUT PROCEDURES

Fryer boil-outs are an essential part of maintaining clean and sanitary processing equipment. The length of time between boil-outs will be determined by production schedules, fryer efficiency, and other factors, but as a general rule fryers should be boiled out at least once per week.

The basic steps are as follows:

1. After the fryer oil has cooled, drain the oil from the fryer (and the make-up tank if necessary).
2. Raise the hood and rinse the fryer, hood and the supply tank with hot water.
3. Fill the fryer with hot water (160°F, if possible) to 1–2 inches above the bundles or heat coils. Add the detergent slowly, making sure it dissolves completely. The concentrated cleaning solution is added per gallon of fryer volume, Spurrier 335-4-2-PP, EcoLab Quadexx, or similar fryer boil-out compound, based on the vendor recommendation and on the total capacity of the fryer, not on the amount of water added to start the boil-out. The temperature of the detergent solution should then be brought up to 190–200°F and permitted to boil for approximately 60–90 minutes.
4. Raise the hood and fill the fryer with warm water to the oil level. Close the hood and raise the temperature of the detergent solution to 190–200°F. Start any mechanical rakes and pumps and run

them another 60–90 minutes.

5. When the boil-out is completed, shut off any rakes and pumps and raise the hood. Drain the detergent solution from the fryer,

6. Inspect the fryer and remove any soil deposits by hand brushing.

7. Fill the fryer with fresh water. Heat the water to 160–190°F and circulate it through the pumps. Add an acidified rinse product to the rinse water, following application levels recommended on the container. This will neutralize any remaining alkaline residuals.

8. Flush the stacks in the hood with warm or hot water. While the fryer is still wet from rinsing, use pH paper to check for any residual alkalinity or acidity. Should any be noted, rinse the fryer with a hose and use pH for a second test. The desired final pH of the rinse solution should be between 6 and 7.5.

TABLE 8.4. RELATIVE MERITS OF CHEMICAL SANITIZERS

The generally recognized advantages and disadvantages of the most popular chemical sanitizers are as follows:

A. Chlorine (application Rate: 200 ppm Chlorine)

Advantages

1. Effective against a wide variety of bacteria; including spores and bacterio-phages.
2. Relatively inexpensive.
3. Not affected by hard water salts.
4. Concentration easily measured by convenient field tests.

Disadvantages

1. Corrosive to metals; hypochlorines more corrosive than organic chlorides.
2. irritating to skin and mucous membranes.
3. Dissipates rapidly from solutions.
4. Effectiveness decreases with increasing pH of most chlorine solutions.
5. Activity decreases rapidly in the presence of organic matter.
6. Odor can be offensive.

B. Iodophors (Application Rate: 25 ppm Iodine)

Advantages

1. Broad spectrum of activity.
2. Visual control by color; forms an amber color in solution.
3. Not affected by hard water soils.
4. Non-corrosive, non-irritating to skin.
5. Prevents film formation due to its acid nature.
6. Activity not lost as rapidly as chlorine in the presence of organic matter.
7. Easily titrated by field methods.
8. Stable; long shelf-life.
9. Spot-free drying.
10. Good penetrating qualities.

Disadvantages

1. Should not be used at temperatures exceeding 120°F.
2. Very slow acting at pH 7.0 or above.
3. Can cause staining problems, particularly on certain plastic surfaces.
4. Less effective against bacterial spores and bacteriophages than chlorine.

C. Quaternary Ammonium Compounds (Application Rate: 200 ppm Quat)

Advantages

1. Relatively non-toxic, odorless, colorless, non-corrosive, non-irritating.
2. Stable to heat and relatively stable in the presence of organic matter.
3. Possess cleaning properties due to its surfactant activity.
4. Eliminates odors.
5. Forms a bacteriostat film.
6. Active against a wide variety of microorganisms.
7. Active over a wide pH range.

Disadvantages

1. Non-compatible with soaps, anionic detergents, and anionic matter in general.
2. Produce foam problems in mechanical operations.
3. Film forming.
4. Not effective against tuberculosis and certain viruses.
5. Forms bacteriostatic film.

D. Acid-Anionic Surfactants (Application Rate: 200 ppm Active Ingredient)

Advantages

1. Non-staining.
2. No objectionable odor.
3. Removes mineral deposits.
4. Effective against a wide spectrum of organisms.
5. Stable in concentrated and in-use dilution form.

Disadvantages

1. Effective at pH only; 1.9 – 2.2 offers optimum activity.
2. Generates foam.
3. Slow activity against spore-forming organisms.